STUDIES IN GENETIC EPISTEMOLOGY

XXIII

EPISTEMOLOGY AND PSYCHOLOGY OF FUNCTIONS

SYNTHESE LIBRARY

MONOGRAPHS ON EPISTEMOLOGY,

LOGIC, METHODOLOGY, PHILOSOPHY OF SCIENCE,

SOCIOLOGY OF SCIENCE AND OF KNOWLEDGE,

AND ON THE MATHEMATICAL METHODS OF

SOCIAL AND BEHAVIORAL SCIENCES

Managing Editor:

JAAKKO HINTIKKA, *Academy of Finland and Stanford University*

Editors:

ROBERT S. COHEN, *Boston University*

DONALD DAVIDSON, *University of Chicago*

GABRIËL NUCHELMANS, *University of Leyden*

WESLEY C. SALMON, *University of Arizona*

VOLUME 83

STUDIES IN GENETIC EPISTEMOLOGY

Published under the direction of
JEAN PIAGET
Professor of the Faculté des Sciences at Geneva

XXIII

EPISTEMOLOGY AND PSYCHOLOGY OF FUNCTIONS

By

JEAN PIAGET, JEAN-BLAISE GRIZE,
ALINA SZEMINSKA, AND VINH BANG

With the Collaboration of

Catherine Fot, Marianne Meylan-Backs, Francine Orsini
Andrula Papert-Christophides, Elsa Schmid-Kitzikis
and Hermine Sinclair

D. REIDEL PUBLISHING COMPANY

DORDRECHT-HOLLAND/BOSTON-U.S.A.

Library of Congress Cataloging in Publication Data

Epistemology and psychology of functions.

    (Studies in genetic epistemology ; 23)   (Synthese
library ; v. 83)
    Translation of Épistémologie et psychologie de la
fonction.
    Includes bibliographical references and indexes.
    1. Cognition in children.   2. Knowledge, Theory of.
3. Functions. I. Piaget, Jean, 1896–   II. Series:
Études d'épistémologie génétique ; 23.
BF723.C5E613   155.4'13   77–6792
ISBN 90–277–0804–5

ÉTUDES D'ÉPISTÉMOLOGIE GÉNÉTIQUE
XXIII
ÉPISTÉMOLOGIE ET PSYCHOLOGIE DE LA FONCTION

First published by Presses Universitaires de France, 1968
Translated from the French by F. Xavier Castellanos and Vivian D. Anderson

Published by D. Reidel Publishing Company,
P.O. Box 17, Dordrecht, Holland

Sold and distributed in the U.S.A., Canada, and Mexico
by D. Reidel Publishing Company, Inc.
Lincoln Building, 160 Old Derby Street, Hingham,
Mass. 02043, U.S.A.

PREFACE

Years ago, prompted by Grize, Apostel and Papert, we undertook the study of functions, but until now we did not properly understand the relations between functions and operations, and their increasing interactions at the level of 'constituted functions'. By contrast, certain recent studies on 'constitutive functions', or preoperatory functional schemes, have convinced us of the existence of a sort of logic of functions (springing from the schemes of actions) which is prior to the logic of operations (drawn from the general and reversible coordinations between actions). This preoperatory 'logic' accounts for the very general, and until now unexplained, primacy of order relations between 4 and 7 years of age, which is natural since functions are ordered dependences and result from oriented 'applications'. And while this 'logic' ends up in a positive manner in formalizable structures, it has gaps or limitations. Psychologically, we are interested in understanding the systematic errors due to this primacy of order, such as the undifferentiation of 'longer' and 'farther', or the non-conservations caused by ordinal estimations (of levels, etc.), as opposed to extensive or metric evaluations. In a sense which is psychologically very real, this preoperatory logic of constitutive functions represents only the first half of operatory logic, if this can be said, and it is reversibility which allows the construction of the other half by completing the initial one-way structures.

Furthermore, with respect to schemes of action, as opposed to general and operatory coordinations, functions constitute the common source of operations and of causality and it is also in regard to this second point of view that the studies contained in this volume present certain new data.

The first part of this work, authored by J. Piaget in collaboration with C. Fot, M. Meylan-Backs, F. Orsini, A. Papert-Christophides, E. Schmid-Kitzikis, H. Sinclair and A. Szeminska, bears above all on constitutive or preoperatory functions and on their gradual transformations into constituted functions, linked to operations. 641 subjects from 3 to 12–13 years of age were tested on this point.

Part II is from the pen of Vinh Bang and relates 5 experiments which he conducted over a period of years on the quantification of constituted functions, and specially on proportionality (based on 353 subjects from 6 to 14 years of age).

Part III comprises two theoretical studies. In the first, J.-B. Grize reviews the history of the logical structure of functions, including the different levels considered in the preceding chapters. In the second, J. Piaget draws the 'General Conclusions' from these studies.

INTRODUCTION

Reading a book by Piaget is like entering a system. To a large extent he and his collaborators (but particularly he) are builders of an impressively structured whole — an experimentally based and controlled set of truth judgements about knowing and knowledge. It seems almost irrelevant to choose just a single book out of his enormous oeuvre. Always one meets a number of fundamental concepts which have arisen through many decades of experimental work with hundreds of collaborators, and from very productive thinking. Without doubt, for more than forty years Piaget has been (and still is) the spearhead and the creating and integrative power behind several generations of famous co-workers in 'the circle of Geneva'. It is impossible to mention all these women and men who were and are explicitly and implicitly present at the fabrication of so many writings. In the highly interesting autobiography of his scientific development, Piaget himself thankfully acknowledges the help of psychologists Bärbel Inhelder, Alina Szeminska, Vinh Bang, Bresson, Greco, and Fraissé, the psycholinguist Sinclair, the mathematicians Henriques and Beth, the physicists Garcia and Halbwachs, the cyberneticians Papert and Cellerier, the methodologist in biology Nowinski, the logicians Apostel, Grize, Wermus, "and so many more".

And here we are at the heart of the Centre International d'Épistémologie Génétique, where the studies in genetic epistemology originate and come about under the direction of Piaget. This particular book from 1968 was the twenty-third to be published since 1956. With the permanent aid of the Rockefeller Foundation since the beginning, and with that of the Fond National Suisse de Recherches Scientifiques since 1964, a production of approximately two volumes per year up till now (1977) has been maintained.

I hardly need to say that this all surpasses the frontiers of psychology in the usual sense. Piaget was and his successors are directing the Faculté de psychologie et sciences de l'éducation of the University of Geneva. From about 1920, as *chef des travaux* at the Institut Jean-Jacques Rousseau under his predecessor Claparède, Piaget began his very original studies in experimental child psychology and especially the cognitive development from childhood to adolescence, in which observations of his own two children played a major role. But significantly (I will come back to the underlying meaning of this) at an earlier date he started as a young man to study 'natural sciences' and wrote a doctoral thesis about the varieties of snail species living under different ecological conditions in Swiss lakes at different altitudes. At the same time he was attracted to and studied traditional philosophy (which, in the course of his scientific development, he surpassed — see his very thoughtful book *Sagesse et illusions de la philosophie*, 1966). This led in his early years, at the Universities of Geneva and Neûchatel, to teaching not only child psychology but also the philosophy of science and the development of ideas as reflected in the history of science. Even courses in human sociology for students (remember the studies on biological ecology) were amongst his tasks.

With these and other starting points in his scientific career, stemming from the 'irrational' roots of his personality (with vital early questions about the essence of life and truth and securities), it is understandable that his very productive experimental and synthetic mind engendered a sizable problem convergency rather early. He wished to discover a kind of 'psychological embryology' of intelligence (using biological preformation) in which the relations between the acting and thinking subject and the objects of his experience, and between the subject and other subjects, had to be seen as a special case of the relation between the biological 'organism' and its surroundings. The difference being in the first case exchange of information and in the second case exchange of material.

Piaget occupied himself temporarily with the development of social cooperation between individuals and groups, with social exchange of satisfactions, with valuation, with motivational aspects, with the growth of normative thinking in morals and justice, but his main line was cognitive development in a more strictly mathematical, geometrical and physical sense and its continuation in the development and the stages of intellectual thinking in science, and the epistemological reflection of science on its own foundations.

To, for example, the Anglo-Saxon reader, this combination of psychology and epistemology may raise doubts. Is this not mixing up too much experimental psychological knowledge with 'philosophy'? Has psychology not to deal experimentally with thinking and has it not to consider the way in which the human mind or the subject constructs images of the objects in the outer world which are, after inner organisation (in which language plays an important role), a true description of reality expressible in words? Piaget and his collaborators have always fought against this kind of reductionism in explaining intelligence and intelligent behaviour. See for example 'The gaps of empiricism' presented in 1968 at the Alpbach Symposium on 'Beyond Reductionism'. Reductionism because action and thinking rather than perception are the main sources of children's knowledge of the world around them and, according to Piaget also as an ultimate result, the adult's knowledge of the principles of logic.

And (to pursue the line of some of his critics further) isn't epistemology a branch of philosophical thinking? Piaget objects strongly to this. He points out that the time of the epistemology of the old masters in philosophy, Plato, Descartes, Leibniz and Kant, is over. There has come instead a modern scientific epistemology, brought about by the reflections of the great scientific minds of modern times. Through their scientific work they came to questions about the foundations of their knowledge and about methods of purchasing truth. This epistemology is far removed from metaphysics and other philosophical fields and strongly connected to the magnificent development of modern (mathematical) logic.

There are many famous names in this connection in mathematics and physics — to mention only one in molecular biology, Jacques Monod. Piaget had an interesting written discussion with Monod on development and evolution and the role of chance, necessity and active self-construction of the living system as a result of perturbation or promotion by its environment and based on regulative and assimilative biological capacities.

Piaget defended himself very early against what he called 'the demon in philosophy' by devoting himself to painstaking experimental work and 'the study of truth' although combined with powerful reasoning. He was always interested in the unobservable to which the 'facts' refer and lead us. In accordance with this and his biological background, already in his early writings he conceived in all domains of life (organic, mental, social) organised or structured wholes and never elements functioning in isolation. Elementary realities we observe are always, in our way of understanding, dependent on bigger totalities which give these elementary realities their meaning. And on all levels there is the problem of the relations between the parts and totality. For example the problem of the species in biology raises the question whether this is 'a reality' or a function of our wider concept in which it is a part. Remember Piaget's studies on varieties of snail species under different ecological circumstances or different totalities of life. This could be considered an analogy of the question of the existence or non-existence of so-called 'mathematical entities' of which we know now that they become meaningless if we develop mathematical thinking into wider, more powerful mathematical structures. And it is again comparable with the historical question in philosophy on 'realism' and 'nominalism'.

And so Piaget turned to our elementary structures of thinking, especially knowing, beginning with and frequently coming back to the study of the very early 'knowledge' of young and very young children. He discovered the 'logic of the child'. Almost from the beginning he started not to be interested in the 'errors' and 'failures' in children's thinking, but in the reason why they must fail in their responses around the problems we 'adults in thinking' put to them. He and his early collaborators developed a fine 'clinical' way of interviewing and interrogating them on their own level, and following them in their stage of logic to higher stages (organisations, systems, structures of thinking, he says) to adolescent and further to adult scientific stages of thinking. Even in the official sciences there is not one but different methods with different stages of thinking behind them. Our knowledge is neither predetermined in the internal structures of our mind nor in the pre-existent properties of things around us. These properties of 'objects' become only gradually known by means of structures of knowledge which are gradually constructed in the developing mind under the influence of experience by action. And here we are in genetic epistemology.

Piaget discovered psychology as an experimental field between biology and the old philosophical analysis of knowledge. He was always in search of experimental knowledge about our growing knowledge. The American Psychological Association, when addressing themselves to Piaget at the occasion of some celebration, said that he had approached philosophical questions in an empirical way and that these studies have led to psychology as a 'by-product'. Yes and no. But this formulation is not precise enough. There is no question about main and secondary activities. It is in essence an experimental and analysing approach to the problem of knowing and the immediate reflection on the developing mental structures behind our growing knowledge. It is better to put it like Piaget himself did: these judgements about knowing trespass upon and go beyond the boundaries of experimental psychology with regard to their meaning, but not with regard to their verification.

After having studied the adult structures of thinking and what he called the mental operations in it, which develop themselves out of regulative activities to ever more complex, finally totally reversible structures of thinking, Piaget turned himself to the most primitive ones, which he found and called 'conservation' or 'constancy' of the 'object' under the many and different situations of perception and use.

The laws that rule this mental development to mental operations in structures of thinking have, according to Piaget, without doubt correspondence with the laws of structuration of the nervous system, which can already partly be expressed in the formalised language of qualitative mathematical structures (groups, lattices, etc.). See for this his important book *Biologie et connaissance*. Piaget puts it as follows: the correspondence of the formal (mathematical, logical) deductive structures to the psychological stages of structuration of mental activity in reality, and with autoregulative, cybernetic models with their realisations in biology, make it understandable that operative thinking (developed out of action of the thinking subject in reality) is the bridge between structured organic life and structured logico-mathematical 'realities'.

Piaget frequently states that no psychological study of our growing cognitive functions is conceivable without the aid of logical or mathematical models, which represent our highest adult forms of cognitive thinking. He states further that no psychological study of our cognitive functions is possible without constant epistemological analysis of what we know.

In the yearly international and interdisciplinary 'plenary' sessions of the Centre International d'Épistémologie Génétique, where the results of very specialised studies are discussed, there is a search for common language, common methods and common conclusions in, for example, highly formalised deduction and experimental verification. In this forum the people mentioned at the beginning of this preface are present, together with important guests from abroad, representing the different disciplines appropriate to the study under discussion. The experiments are analysed as well as the theory that lies behind them. Different interpretations of the various disciplines with regard to the same experiments are welcomed as a main road to higher reciprocal understanding and higher forms of knowing, and also as a starting point for further questions and further experiments.

The specific importance of this book lies in the fact that it raised the question and opened discussion on something that had been dealt with only slightly in the previous years of study on the genesis, stages, and role of mental operations in cognitive thinking. Formerly, up to the age of 6 the child was considered to have a sensory-motor intelligence, but till about that age no operational intelligence, be it only mental operations in the sense of handling the representations of concrete things. During that period there is already an intelligent behaviour, even before the appearance of language, but only in a sensory-motor sense, in dealing with concrete things. During that period there are 'schemes of actions' with coordination and organisation of adaptive action. But this book presents a much more positive view of this period from, say, 3½ to 6. Here again it is with the aid of mathematical, formalised models of adult thinking that the origins of this kind of thinking are studied in the psychological reality of the child and a beginning of understanding is founded on the relations between thinking in functions, which is essentially thinking in dependencies, and operational intelligence, with reversibilities and the possibility of transformations within a structured whole. The function is considered as a precursor of the mental operation.

Also the origin of functions is discussed, whether in the physical world around the child or in his actions, his own handling of objects. The book speaks abundantly for itself.

Finally, coming back once again to the wholeness of all the works in and about the 'circle of Geneva' around Piaget, I may mention here the existence of a *Catalogue des Archives Jean Piaget* of the University of Geneva under direction of Professor Bärbel Inhelder. This bibliography is divided into three parts, one of which contains all the works of Piaget from 1907 till now, with 1500 titles. The second part contains about 800 index cards dealing with all the works of immediate collaborators. The third part contains the so-called secondary literature engendered by Piagetian thinking with about 1750 index cards. The publisher is G. K. Hall and Co. Publications, Boston, U.S.A.

Besides being a model of modern scientific world citizenship, with a never-ending inquisitive, experimental attitude, Piaget is for many a kind of ideal of the old European mind. His unbreakable tie with nature, to which belong periodic retreats and walks in the Swiss mountains, is combined with his powerful rational, integrative thinking and reasoning, far above the fortuitous situation; his healthy taste for good companionship and good life is combined with charming simple style and manners and with an outside soberness; and even his very human slyness. These are all aspects of a kind of Olympianship.

A. SUNIER

TABLE OF CONTENTS

PART III / THEORETICAL PROBLEMS

PART I

FROM CONSTITUTIVE FUNCTIONS
TO CONSTITUTED FUNCTIONS

CHAPTER 1

THE COORDINATION OF PAIRS [1]

We have observed for some time[2] that when children are asked to order a series of objects in increasing sizes $A < B < C$, the youngest subjects simply proceed in pairs, such as CF, AD, EH, etc., maintaining the order 'a little one, a big one', but they are not subsequently able to coordinate them into a single series. This also occurs with the first classifications made from 'figural collections', with the youngest subjects doing well in constructing a composite figure, such as a row, but in general placing an object in relation to the one just before it (e.g., a square after a square, or a sheep following a shepherd), without however, taking into account the other preceding terms, in other words, once again without coordinating the pairs in spite of the linear form of the whole. In 1947, Wallon (*Les origines de la pensée chez l'enfant*) focused on this idea of pairs seeing in it the most elementary form of cognitive structuring. This idea was also contained in Hoeffding's comparison (in *La pensée humaine*) of the workings of thought to the successive positions of a compass, where one of the legs, although placed on a given point, provides no information until the other leg is placed on a second point or on another object.

If we define functions mathematically as univocal relations towards the right,[3] i.e. as ordered pairs, then the pairs just referred to above, however elementary they may be, already constitute functions. But if in addition functions are considered, from a psychological standpoint, as the expressions of the schemes of assimilation of actions, then functions are already present in the conceptualization of any action which modifies an object x into x' or y, thereby also constituting an ordered pair (x, x') or (x, y). Since we hypothesize that functions constitute the common source of operations and causal systems, it must therefore be possible, even in the most elementary cases, to find such (x, y) pairs prior to the differentiation of operatory and causal systems. (These pairs are then susceptible to orientation in either of these two directions.) In the following research we will take such (x, y) pairs in general as a substitution of y for x. This substitution[4] can be conceived as the product of an action (or operation) of the subject (e.g., selecting y starting from x and finding a certain transformational correspondence between x and y) or of a causal action (modifying x into y by making it larger, changing its color, etc.) or even of a simple movement (displacing a movable object by substituting position y for the initial position x).

Such pairs produced by a substitutive action bring up the problem of their composition among themselves; in other words, of the stages of their coordination. Is it possible to coordinate these pairs simply because they have been constructed, or does there exist a level of development at which they cannot be so coordinated (as we recalled above regarding seriation, etc.)? If so, is it possible to assign to pairs a certain structure sufficing (provisionally) unto itself and having, for example, some analogy to 'category' as viewed from the standpoint of the mathematician (MacLane, etc.)?

We will use two experimental tasks which are isomorphic except in their concrete manifestations. One is composed of flowers which can be substituted for each other according to two different sizes and colors. The other consists of segmented trajectories and an object which moves along them as one position is substituted for another. In both cases the subjects are asked to reach certain goals by combining two or three successive substitutions, thus keeping the problem as 'childishly' simple as possible. The experiment shows that although the possible compositions are sometimes constructed immediately, this does not occur as frequently as might have been expected, for these compositions are often made by successive trial and error. This would seem to point to the fact that the existence or formation of pairs precedes their coordination and that a pair does in fact constitute a sort of unit or unit-structure, which if not independent at least leads to a relatively independent awareness. Furthermore, starting from such a point of origin (which is at least relative, since actions can be coordinated well before they are conceptualized or the subject is aware of them), functions can be oriented in different directions through the development of either classes or systems of relations, or causal systems or laws.

1. EXPERIMENT I: THE SUBSTITUTION OF FLOWERS

The child is given 4 charts, each having 8 boxes (Figure 1). The upper boxes of all the charts contain, from left to right, a large red flower, a small red one, a large blue one and a small blue one. In chart I, the same order of flowers is duplicated in the lower boxes, i.e., when an upper element is substituted for the one below it, no change occurs. In chart S each of the upper flowers corresponds to the one below it in color but not in size. In chart C, for each upper flower the corresponding lower one is of the same size but of a different color. Lastly, on chart D the upper flowers have a term-by-term correspondence to the flowers of the lower boxes such that the flowers differ in both color and size. The flowers are painted on small cards, some of which, in addition to those in the boxes, are kept aside.

Once we are certain that these correspondences have been well understood by the

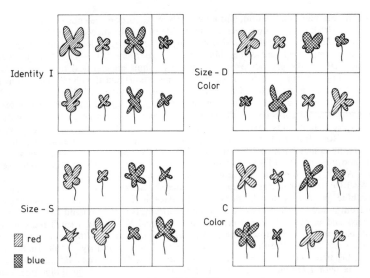

Identity I

Size - D
Color

Size - S

🗆 red
🗆 blue

C
Color

Fig. 1

subject, a game of exchange is suggested. The rules are as follows: whenever the child puts any one flower on another flower of the same size and color which is found in one of the upper boxes of charts *I* through *D*, he will have the right to take in exchange the corresponding flower found in the box directly underneath. The subject is given the chance to make a few preliminary practice plays until he has sufficiently grasped the rules of the game. This done, he proceeds with the actual experiment which consists of obtaining, for a given flower (e.g., a large red one), another flower (e.g., a large blue one). However, when the 'store' for direct exchange is closed (in this case chart *C*), the substitutions must be made in at least two exchanges (here *S* and *D*). Testing begins with the closing of chart *D* and ends with the closing of chart *I*. In some cases (Technique II) the experiment begins with the child constructing the charts himself and in other cases (Technique III), after the construction, the chart is covered by a screen bearing only a label stating the nature of the possible exchange: color, size, both of these, and identity.[5]

The first results of this experiment show that young subjects experience remarkable difficulty in composing the pairs of exchange or substitution when proceeding with anticipations or directed compositions, in contrast to the ease with which they understand these compositions once they have been successfully attained by trial and error or have been indirectly suggested:

CLA (Age 5): 'If you must find a big red one with a little blue one which chart will you use? — *Here* (*D*: correct). — Try to find the same flower but this time don't use this chart again (*D*). — (He looks carefully for the desired pair on the three other charts): *I can't do it.* — Try to do it in several steps, try to start somewhere. — (Long hesitation: he takes the big blue one). Good, and now? – (New hesitation: he ends up by changing the color.) – Alright, now start over (same problem). – (He succeeds right away): *I first took a big blue one* (shows *S*) *and then a big red one* (*C*). – Good. And if I give you a little blue one, find me a big blue one. – *I go there* (*S*). – Yes, but the store is closed. – (He looks and chooses *I*, which gives him an identical one): *That doesn't work*. – Try as before. – (He takes a big red one from *D* and then he looks until he finds what he wants on *C*.) – Now I give you a big red one, look for a big blue one. – *There* (*S*). – Yes, but it's closed. – *Then I have to look for a big red one below and a big blue one above* (he then inverts the direction of the pair, which is progress in mobility, but doesn't see that it leads back to chart *C* which is closed; he finally succeeds by trial and error). – Now, find a little red one with another little red one. (After several unfruitful detours, he comes back to *I* but it is closed). – *Like that* (two exchanges on *S*). – Yes, and another way? – (Two exchanges on *C*). – And again? – (Two exchanges on *D*). – And if you don't use the same chart? – (Attempts): *I can't do it*.

NAT (6; 0) – also declares '*I can't do it*' when she is asked to get the small red one with a big blue one if *D* is closed. She is made to place her flower on *C*, she takes a small blue one from it and then succeeds on *S*. When she attempts it for a second time, she succeeds by trial and error, but after going through *I*, *C* and *S*. To find a small red one with a small blue one, *C* being closed, she tries *D*: '*Here, but there is a little red one above and a big blue one below, I can't do it*. – Take the big blue one and try. – (She finds it on *S*). And a big blue one with a big blue one (*I* being closed)? – *I can't do it: there's no big blue one with a big blue one*. – Try like before. – (She takes a small blue one from *S*): *No, that won't go here* (then continuing with her exploration she comes back to *S* and sees that with the small blue one she can find another large blue one). – And here (*C*)? – *That doesn't work*'. Etc.

PAS (6; 11) who constructed the four charts including *D* by himself exhibits the same non-compositional trial and error and after the fact comprehensional reactions, but makes no generalizations other than *CC = I*, etc.

The difficulty of these compositions cannot be attributed to a failure to comprehend the instructions for we will find the same lack of coordination in the case of the trains where the joining of two segments representing a transfer between railroad lines would seem much more intuitive. This difficulty therefore requires a more general explanation which can be sought in two directions. The first is causal in nature and can be ascribed to the transitivity of an instrumental action: except in the simple case where a thrust is transmitted to an intermediate object (stick, etc.) which in turn conveys it to a movable object not directly touched by the body itself, it is sufficiently well-known that a child under age 7 cannot compose the desired means (hooks, etc.) for attaining a material end. The second involves operatory transitivity: until about age 7, the child cannot conclude that $A < C$ or $A = C$ if $A < B$ and $B < C$ or $A = B$ and $B = C$, and even when he does establish the product of this composition, he still does not draw the required inferential conclusion. In the case of these particular compositions, the substitution of *C* for *A* through *C* for *B* and *B* for *A* results in a situation which is mid-way between instrumental transitivity and logical transitivity, wherein lies the initial difficulty.

However, it seems clear that while a single pair already constitutes a structure prior to its composition with another, this composition, as soon as it is observed to be possible, even when not anticipated, is understood as an extension of the original pair, thereby making possible a generalization of the model. It is true that as yet no deductive generalization has been made at this level, in the sense that since one of the two solutions was found empirically (for example C then S to find D) the symmetrical solution (order: S then C) was not deduced nor even sought empirically, as if there existed only one possible order in the succession of the substitutions. This lack of reversibility (which here refers to the reversibility of the order, i.e. commutativity) is, however, natural at the preoperatory level and, moreover, does not extend to all of the functions in question. On the contrary, it is remarkable that as soon as CLA (age 5) discovers the possibility of obtaining the identity I through two successive inverse exchanges on chart S (small red one → large red one, then the inverse) where the two inverse pairs are placed next to each other, he immediately generalizes this solution for charts C and D. The same is also done by other subjects aged 5–6.

In the next stage, observed from age 8 on, the discovery of the composition of the substitutions no longer occurs solely through empirical trial and error, except at times in the first attempt, but also by deduction from looking at the charts. True, the subject still needs to look at the charts and cannot merely rely on evoking them by name and operatory significance as he will be able to do in stage III, but by looking at them he is able to mentally combine the possible substitutions. In other words, during stages I, II and III, actions are progressively interiorized while the structure of their scheme and of the functions which express it remains the same. It is precisely this functional structure which we will seek to isolate from the standpoint of 'functions', in the abstract sense of the term, as we are presently doing from the standpoint of psychological 'functioning'. Let us also note that if the next level of progress is to be marked by the attainment of commutativity, the subjects observed have not yet attained it with Technique III and need a factual control in order to be assured of its validity:

DAN (8; 9): 'I will give you the little blue one and you look for the big red one. – *Here* (*D*). – Yes, but if the store is closed? – *I can't do it.* – And with a little detour? – *I can't do it. Yes I can, here* (*C*) *then there* (*S*). *That's it.* – And if I give you the little red one to find the big blue one? – (He looks). – Can it be done in one step if (*D*) is closed? – *It can't be done* (he finds *S* then *C*). – We will now close *S*, you find the little blue one with the big blue one. – *Here* (*C* then *D*). – Is there another way to do it? – *Yes, there* (*D*) *and there* (*C*). – What is the difference? – *First of all, I did* (*D, C*), *then* (*C, D*). *This time I did it in the opposite direction from before.* – If (*C*) is closed find the big blue one with the big red one. – *Like that* (*C, D*). – How many charts? – *Two.* – Is there another way? – *Yes (D, C). I did the opposite.* – Look for the little red one with a little red one. – *Here (I).* – And, if it's closed? – *Like that (D, D).* – And otherwise? – *Like that (S, S).* – And another way? – *Here (C, C).* – And another way with several charts? – (He finds *S, C* and *D*) How many? – *Three'.*

YVE (8; 10, Technique III) constructs the four charts including *I* after some trial|and error and labels the papers used to cover them as follows: 'color (*C*), alike (*A*), size (*S*) color and size (*D*)'. He is asked to find a big red one with a little blue one: *'Here (D).* – And if it's closed, can you do it another way? – No. – If your path is barred on your way to school, what do you do? – *Take a detour.* – And here? – *On (S) I could* . . . – Get what? – *A big blue one.* – And with that? – *A big red one on (C).* – And to find a little red one with a big blue one? – *Go there (C), I take the little blue one* (wrong). – What does (*C*) give you? – *Color. It gives me the big blue one and I go to (S), it gives me the little blue one.* – Find the little blue one with the big one (*S* closed). – *I go to (I), it gives me the big blue one, from (C) the big red one and from (D) the little blue one.* – Can you do it another way? – *No.* – And if you began by (*D*)? – *Oh yes, it gives me the little red one and (C) the little blue one'.* The same thing happens with *C* closed, he finds a solution but not the inverse. With *I* he finds a solution and its inverse but, until the screen has been lifted, he cannot see that a double exchange e.g., (*D, D*) is possible on a single chart.

We can thus see that, from age 8 on, all solutions including their reciprocals (commutativity) can be found without trial and error when the charts are visible. But when they are covered by screens, although the subjects can find one solution, they fail to do so for its inverse, even with a double substitution on a single chart. In stage III, on the other hand, the deductive (and hypothetico-deductive) processes at the subject's command make it possible for him to solve these problems even without a perceptual inspection of the charts (Technique III):

ARI (10; 1, Technique III) constructs the four charts and names them *'color', 'size',* *'both' (D)* and *'just the same' (I). D* being closed, she goes through *S* then *C*: 'Could we go first through (*C*) then (*S*)? – *'No'*, then changes her mind and finds it. *C* and then *S* being closed, she finds the solution and its inverse each time. In order to find a little red one with another little red one with *I* closed, she designates (*D*): *'I get a big blue one, then at (S) a little one, and I go to (C) and I find the little red one.* – Can you do it another way? – (She indicates another order). – And with (*C*) or (*D*) alone? – *No, I can't do it.* – Why? – (Thinks deeply while murmuring): *Yes, I do it twice. Here (D) I get a big blue one, then I redo it and I get a little red one.* – And with (*C*) only? – *Yes, I can also do it twice.* – And with (*S*)? – *Twice again'.*

All of the questions can be solved at this stage even when the subject is not looking at the charts. Let us repeat, however, that the structure thus found was discovered as early as stage I, albeit by trial and error and by comprehension after the fact. Nevertheless since the pairs involved already link terms of exchange or substitution which are readily understood as such and since the initial trial and error only occurs when two or more of these pairs are composed, there is continuity between the structure of the pairs themselves and that of their compositions. Before attempting to isolate the general characteristics of these compositions, let us also examine the reactions to a second experiment which is functionally isomorphic to the preceding one.

2. EXPERIMENT II: THE SUBSTITUTION OF POSITIONS

In the present experiment the subject had to deal with four kinds of flowers. Let us call the flower we start with *1* and the flowers we must end up with *2*, *3* and *4*. The paths connecting them may be simple $(1 \rightarrow 2, 2 \rightarrow 3,$ etc.) or compound $(1 \rightarrow 2) + (2 \rightarrow 3)$ or even short-circuited $(1 \rightarrow 3$ and $2 \rightarrow 4)$ with the object of the task being their composition. Let us now replace these flowers by positions 1, 2, 3, 4, (Figure 2) which are linked together by green $(1{-}2$ and $3{-}4)$, blue $(2{-}3$ and $1{-}4)$ or red $(1{-}3$ and $2{-}4)$ lines and let us call the changes or substitutions of position 'displacements'.[6] The problem of the composition of pairs can then be posed to the subject in terms of displacements. He is shown Figure 2 which is 12 x 15 cm, and is told that the colored lines are railroad tracks on which a train (= small rectangle of the same color as the line) runs.

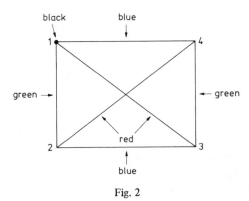

Fig. 2

Each train can only travel along one straight segment and therefore, to go from, for example, 1 to 3, it will be necessary to change trains at 2. Furthermore, some lines will be 'closed' from time to time so that when the green lines are closed, to go from 1 to 2, it will be necessary to use routes $(1, 3) + (3, 2)$ or $(1, 4) + (4, 2)$. Finally, all departures are made from 1.

It is clear that the problem posed is exactly the same as the one in the preceding experiment: each $(1, 2)$ pair represents a substitution of positions and the coordination of the pairs (for example $(1, 2) + (2, 3)$), consists of a composition of two substitutions which yield a univocal resultant (here $(1, 3)$). The only difference is that, intuitively, this composition seems simpler since it is only a matter of changing trains and joining the trips or vectors end-to-end. The three questions systematically asked were as follows: (1) to go from 1 to 4 with the blue routes closed; (2) to go from 1 to 3 with the red routes closed; and (3) to go from 1 to 2 with the green routes closed. In addition, the subjects were always asked to find a second solution and often a return route. Lastly, the subject was asked to indicate a trip in which point 1 would be the point of departure as well as arrival.

Regardless of how easy the composition of pairs by substitution might seem, we noted that they resulted in difficulties which were quite comparable to the preceding ones. In making a table of the correct answers to questions (1) to (3), we found the following to be the case as a function of age, listing separately single successes (one correct solution) and double ones (two correct solutions, the second being the reciprocal of the first, e.g., $(1, 3) + (3, 4)$ for $(1, 2) + (2, 4)$ or red + green for green + red). The table indicates the percentage of correct answers in proportion to the number of questions:[7]

Age	4–5 yrs.	6 yrs.	7 yrs.	8 yrs.
One solution	52	78	96	100
Two solutions	16	64	66	86

We thus see that from age 6 on, three-fourths of the subjects are successful in finding at least one of the two possible solutions and that from 8 on, they are successful in finding both since the givens of the problem remained perceptible and there was no use made of formal deduction by hiding the chart as in Technique III of the preceding experiment: successes are at the same level for the same thought processes. We can also see that the successes are in no way immediate and that at 4–5 years of age, we find as before incomplete solutions: 84% of the questions remain without a second solution and 48% lead only to the construction of pairs without sufficient composition among themselves.

It is this inability to make compositions which we must examine and which corresponds to the failure to coordinate the pairs in Experiment 1. Five types of errors can be distinguished in this respect. The first type is only found among 4 year-olds and in one or two 5 year-and-a-few-month-olds. It consists of supplying only one pair determined by point of departure 1 (Figure 3 I). The second type, found in subjects aged 4 to 8, also consists of giving only one route centered on the point of arrival without taking into account the point of departure (Figure 3 II). The third (up to age 6), consists of providing two pairs or paths which have a common origin but no common

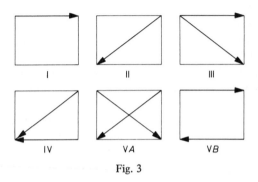

Fig. 3

point of arrival (Figure 3 III). In the fourth type, the lack of coordination is marked by two points of origin which nonetheless converge at the specific point of arrival requested (ages 4 to 6: Figure 3 IV). The fifth and last type (ages 4–8) is the most frequent together with type II, and the most interesting: it consists of two unlinked paths one of which correctly corresponds to point of departure 1 with the other corresponding to the specific point of arrival requested (point 2 in the case of Figure 3 V A and V B). But it should be noted that in general these types are mixed for a given subject and with the exception of two 4 year-olds, all of the subjects provided at least one correct answer:

ALA (4; 6) understands the instructions well and is right away able to construct trips such as (1, 2) then (2, 4) and (4, 3), etc. On the other hand, when he has to compose an itinerary with detours, he gives a mixture of correct answers and errors of the I, II, III, IV and V types, the correct answers consisting of only one of the two possible solutions.
    For example, to go from 1 to 4 without the blue lines, he finds (1, 2) + (2, 4) which is correct. But, to go from 1 to 2 without the green ones, he only finds (1, 3) without adding (3, 2) (type I error). To go from 1 to 3 without the red ones, he gives (2, 3) without (1, 2) (type II), etc.
    BUR (5; 6) also makes correct spontaneous compositions, yet when it is her turn to answer the questions indicated, she thrice makes type II errors: to go from 1 to 2 without using the green lines, she gives for example, a red (4, 2) line: 'You started out from here (1)? – No. – Then how can you do it? – I can't do it'. – She is shown how, but she continues to have the same reaction in other instances. She finally ends up by

making a type V error: to go again from 1 to 2, she gives (1, 3) and (4, 2) without coordinating these two vectors. On the other hand, she does react correctly to one of the questions asked and even finds the reciprocal of her solution.

We note once again that these compositional failures have been controlled in three different ways. First, when the child is asked to find the return route, the errors made are the same as those made for the initial trip and, for a given subject, are generally of the same type. Secondly, when we were not sure certain subjects had fully understood the instructions, a passenger was added to the trains who was to be transferred from one train to another whenever a change was made. The errors made in this case were of the same types as before. Finally, to determine if the subjects were being misled by color, we used a model that had six different colors for the six different routes, but this brought about no change in the reactions.

With regard to round trips, subjects aged 4–6 mostly gave immediate inversions (such as (1, 2) and (2, 1)), but also some compositions, these becoming almost general at 7–8 years of age.

In summary, as with the problem of the substitution of flowers, the child sometimes stops at isolated pairs before being able to make compositions the difficulty of which derives from the fact that they involve a sort of intermediate transitivity between instrumental transitivity and logical transitivity. Correct solutions become possible from ages 4–5 on and become understood as soon as they are discovered by trial and error.

3. CONCLUSION: THE STRUCTURE OF 'CATEGORIES'

In accordance with the hypothesis that functions express the links proper to schemes of actions, it must be possible to find functional structures which precede operatory structures since actions can be physically coordinated before operations drawn from the most general coordinations are constituted. The two preceding experiments demonstrate a very early formation of 'pairs' and an equally early composition of same (with successes not general, but reaching 50% from ages 4–5). The flower experiment consists of simple exchanges and thus represents readily understandable everyday behavior. In the second experiment, the system involved can take on the operatory form of a 'group of displacements', yet, as we must point out, this system first appears as early as the sensorimotor period (12 to 18 months)[8] in a specifically empirical form through the step-by-step coordination of actions, prior to being structured by deductive operations around age 7–8.

We will therefore assume that preoperatory functional structures do exist and that we must now find a model for them. Mathematicians, after having isolated the major structures of groups, networks, etc., remarkable in their elegance as well as in the richness and high degree of coherence of their internal compositions, have begun to search for more elementary and even more general systems. Starting out from the idea that the simplest functions are 'applications', sources of multiple morphisms, they have attempted to close the gap between what S. Papert so rightly calls the operations 'of the mathematician' as opposed to those 'of mathematics'. This is tantamount to emphasizing the role of limited but effective actions which are carried out irrespective of the situation. As a result they have been able to isolate systems which are remarkably fecund but which are of no concern to us here. What does interest us, from the genetic standpoint, is the fact that some of those systems, which are by their very nature elementary, are likely to be of technical as well as trivial significance and could consequently be of use to the psychology of intelligence or to genetic epistemology as well as to mathematics.

Such is the notion of 'category' developed by Mac Lane, among others, and reviewed by J. B. Grize in Chapter 14 of this volume. In very general terms, a category is a set of objects which includes the functions linking their properties (this accentuates the latter more than the objects and, as Grize emphasizes, thereby permits a position mid-way between extension and intension, a position which is very useful from the genetic point of view). It differs from an operatory system in that the composition involved – which, when at all possible, is associative – is not defined everywhere, whereas this would not apply to a closed extensional operatory structure. (Note, however, the nature, intermediate in this respect, of the 'groupings' for which compositions are not possible other than in a 'contiguous' fashion, i.e. step-by-step). Furthermore, since functions are oriented, there exist two neutral elements, one to the left and one to the right.

Grize's formalization of 'category' suggests that it can incorporate not only the composition of two pairs, but also (and this is essential in the light of the preceding results) the pair (a, b) itself in that it comprises the functions (a, a), (a, b), (b, a), and (b, b). If this then is the case, we can conceive of the progressive enlargement of such partial systems, limited at first by the nature of the possible actions, but leading step-by-step from functions to operations through the interplay of coordinations added one to the other until a complete and above all reversible system has been constituted.

Finally, the use of the term 'constitutive functions' has real benefits, not only because it preserves the continuity between functions and operations without reducing the latter to the former, but also because it makes possible a functional analysis of physical actions whose irreversibility renders them irreducible to operations. It is the attribution of the latter to objects which ends up by completing the external functional links until the causal explanations which derive from the system *qua* system are attained.

NOTES

[1] With the collaboration of C. Fot.

[2] Piaget and Szeminska, *The Child's Conception of Number*, Routledge & Kegan Paul, London, 1952.

[3] *Translator's Note*: The phrase 'univocal towards the right' ('*univoque à droite*') refers to the property which distinguishes functions from relations, i.e., a function is comprised by a set of ordered pairs (x, y) such that for any given x from a set X, the function f assigns only one value (y which is a member of the set Y) to that x. Thus we write $y = f(x)$. If additionally, the uniqueness condition is fulfilled in the other direction, i.e., if for each y there is only one x, then Piaget would say that the function f is 'bi-univocal' or 'one-to-one'.

[4] In his theoretical chapter (14), Grize proposes with reason that constitutive functions be considered as articulated, not by operations (in the sense of groupings and reversibility), but by the 'combinators' found in the very general combinatory logic of Curry, etc. Now one of these fundamental combinators is the 'permutator' C, which is specifically utilized in this chapter for the construction of elementary pairs. We only note that one can conceive of varieties or degrees with regard to this permutator. Thus we can write C_0 if a is simply 'substituted' for b; C_1 if there is a reciprocal substitution such that ab is permuted to ba, etc. Furthermore, besides the permutator C which relates to the actions of the subject, we can conceive of a permutator C' relative to the object. Thus in this case, the sequence ab, then ba would signify a displacement of a with respect to b. It is through this displacement or substitution of positions that movement is first of all conceived, ordinally and prior to the construction of any metric bearing on the sizes of the displacements. This notion will be utilized in §2 of this chapter.

[5] The label is formulated by the child himself.

[6] From this point on, the route $1 \rightarrow 2$ will be designated by '$(1, 2)$', etc. (See Figure 2.)

[7] Twelve 4–5 year-olds answered 36 questions; 14 6 year-olds – 42 questions; nine 7 year-olds – 27 questions; ten 8 year-olds – 30 questions. These subjects are different from those questioned in §1.

[8] Piaget, J., *The Construction of Reality in the Child*, Basic Books, New York, 1954.

CHAPTER 2

FROM CONSTITUTIVE FUNCTIONS TO EQUIVALENCE CLASSES[1]

When children set up classifications prior to the constitution of classes based solely on similarities and differences, i.e., on a system of objective equivalences, we observe an initial stage in which there exist only 'figural collections'[2] where the following principle applies: not only are the elements of a class spatially arranged in such a way as to give the latter an overall shape (row, rectangle, etc.), but also (at least in the case of simpler forms) in such a way that one element is linked to another by reason of various 'suitabilities' ('*convenances*') which are unrelated to similarity. For example, a triangle is placed on a square in order to make a house and its roof, a nail with a hammer, a fir tree with a hut (instead of with another fir tree), etc. We also know that the early definitions made by children are not made on the basis of 'kind and specific difference' (last stage), but rather by 'usage' as evidenced by the use (in all languages) of the words 'it's for': a mountain 'is for climbing', a snail 'is for crushing', etc. It would therefore seem clear that prior to operatory classifications based on additive class inclusions in extensions and on objective equivalences of different orders in intension, there exists a mode of classification based on the relationship between actions which are functional in two ways, i.e. as the applications of schemes of actions and as the expressions of dependences.

In order to study constitutive functions, we must therefore specify the direction of these kinds of primitive 'applications' and above all study how the subject passes from functional links based on 'suitabilities' or concrete dependences (spatial, causal, finalist, etc.) to equivalence classes based on objective similarities. We thus had to find a situation where the elements to be classified could each in turn be considered as instruments of an action, classifiable according to the schemes of these actions, and as objects with different properties such that they could be grouped according to their qualitative or quantitative correspondences. It was above all important for these actions and these properties to be of the same nature (i.e. both spatial) so that there would be a sufficiently continuous progression between centration on the action and decentration on the object and such that it would thus be possible to understand the reasons for the passage from the initial constitutive functions, by concrete dependences, to functions constituted by objective equivalence classes. Let us recall that the term constitutive function refers to those links or dependences which are inherent in schemes of action at a preoperatory level. These functions represent the point of origin, whether of operations which are properly of the subject (resulting in constituted functions which, in this particular case, would lead to the construction of equivalence class inclusions), or of causal systems at a level where causality consists of operations attributed to the object (which will not enter into the present situations).

In this experiment the child is asked to cover the white part of square base cards with movable red cutouts of varying shapes. Since several of these small cutouts can be used to cover a given base in spite of their very distinct shapes, it will be possible to determine the degree to which these are analyzed according to the 'suitability' of the action or according to the properties (similarities and differences) of the objects themselves. Since these properties already intervene in the action (the 'suitabilities' depend on them), although much more loosely (since, let us repeat, several of the movable cutouts can fulfill the same condition of 'suitability'), we will thus find equivalence classes at all levels, albeit very different in form. At the level where actions are primary, the only equivalences possible are those resulting from a substitution within a given action, while at the level where objects are primary, equivalences are based on the similarities and differences existing among the objects. From that point on, the 'suitability' to the base card intervenes as a generic framework and no longer as the specific characteristic which overrides the similarities themselves. As we can see, this situation is very adequate for the purposes of the question we are addressing, i.e. studying the conditions underlying the passage from constitutive functions and the dependences or 'suitabilities' deriving from actions, to objective equivalence classes.

1. PROCEDURE AND GENERAL RESULTS

The materials consist first of all of 4 stationary base cards which we shall call A, B, C and D (Figure 4). The upper part of these cards is red while the lower part is white. The task is to cover the white part with movable red cutouts such that the bases A through D are entirely red. The movable cards will be designated A_1, A_2, A_3, B_1, B_2, etc. in accordance with their 'suitability' to the corresponding base card. Figure 5 shows the cards C_1, C_2 and C_3 corresponding to base C (10 of the 12 movable cutouts are completely distinct and two are identical to two of the other ten; all are|laminated| onto plastic squares thus making it possible to apply them more easily onto the bases). There are also some movable white cutouts (also laminated, specially since their parts are not

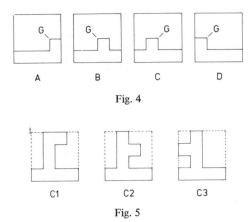

Fig. 4

Fig. 5

always contiguous), which we shall call A'_1, or A'_2, etc. and which represent the complementary surfaces of A_1, A_2, etc. By adjusting one of these movable white cutouts, for example C'_1 (Figure 6) onto its complement C_1, we thus obtain a square which is red and white without any overlapping.

Let us further specify that each of the movable red cutouts A_1, A_2, etc. can suit only one of the bases $A, B \ldots D$ (exclusive 'suitability'). Additionally, we have two red movable cutouts E_1 and E_2 which can cover the white parts of more than one base, i.e. E_1 can cover bases A and C and E_2 can cover bases B and D (non-exclusive 'suitability').

C'1   A

Fig. 6 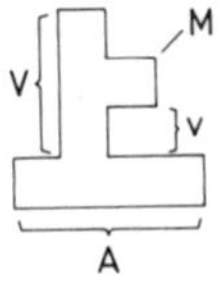   Fig. 7

Lastly let us point out that, for the analysis of the parts of a cutout, we will use the symbols indicated on Figure 7: M for the little segment of variable position, H for the large horizontal segment common to all of the figures, V for the large vertical segment and v for the part of V between H and M. Likewise, for the bases $A \ldots D$ (Figure 4), we will call G the small white square of variable position which determines the choice of the red movable elements by correspondence to V.

The experiment begins with the presentation of the bases $A \ldots D$ and of the 12 movable elements A'_1, $A'_2 \ldots D'_3$ (excluding the E's), which are mixed and distributed on the table. The child is then asked to discover the 'suitability' of each of the elements to its base and then to say (by anticipation and, if he can't, by actual attempts) whether any one of them can 'suit' two bases (exclusive 'suitability').

We then ask a second group of questions: (1) Indicate the similarities and differences existing between the elements of a given base: A_1 compared to A_2 and to A_3, etc.; (2)

Idem between the elements of different bases (A_1 to B_3, etc.); (3) Indicate whether certain characteristics of the elements 'count more than others': for example M or V as compared to H, etc.; (4) Consider whether two elements are: (a) identical (there are some for B); (b) very close in shape; (c) equivalent in terms of 'suitability'.

The third problem is to find the inherent characteristic of an element E, introduced at that time (non-exclusive 'suitability').

In the fourth question, we ask for reactions to the complementary elements ($A'_1 \ldots D'_3$) following a procedure identical to the one used in the first and second questions.

After having placed across from each of the bases $A-D$, their respective elements (A_{1-3}, etc.) and their complements (A'_{1-3}, etc.) we then ask the fifth question regarding the relations: (a) between a base and its elements; (b) between the elements and their complements; and (c) between a base ± its red elements and its complements.

Lastly, having removed the bases, we ask the subjects to classify the elements as follows: 'put together everything that goes well together'. This same classification is also asked of subjects who have not been asked the first five questions.

A total of 50 subjects were tested: 11 aged 5 to 7, 16 aged 8, 14 aged 9 to 10, 14 aged 11 to 13 and 4 older. The levels observed can be broken down broadly as follows:

In the first stage, classification depends essentially on the actions of the subject, objects being considered equivalent to the degree that they are assimilated into the same scheme, thus integrated into the same action: whether they be real actions, such as when the subjects cover a base card, or whether they be partially fictional (this serves to show how little decentration there is on the object), such as when the subject considers two red elements to be equivalent because in order to make them equal, it would suffice to displace one of their segments, for example M. This general subordination to actions can even override perceptual similarities.

In the second stage, classification is based on the characteristics of the object in terms of similarities and differences: i.e. sizes, positions, distances between the parts and the relative importance of some parts over others. These comparisons are made element by element irrespective of their exclusive 'suitability' to any given base, but rather as one characteristic among others. It is evident that so far there are no class inclusions nor included subclasses.

In the third stage, the subject takes into account all objective similarities and differences using the exclusive correspondence of an element to its base as a generic or hierarchical framework. Due however to the successive positions of the element V in relation to H, the subclasses are distinguished by the differentiating properties of their terms in relation to V.

We will proceed below by groups of questions and indicate for each group the evolution observed.

2. EXPLORATION AND EXCLUSIVE 'SUITABILITY'

(The elements A_{1-3}, B_{1-3}, ...)

While it might appear that the material successes achieved (discovery of the suitabilities by covering and above all of their exclusive nature) are of no interest in terms of the study of functions as such and could only be of significance to the psychology of intelligence, the methods used by the subjects to arrive at the correct response, as all did, already in fact reveal the types of links established by the subjects. In general we can discern two basic patterns, i.e. either the child proceeds by successive trial and error without anticipation nor generalization, in which case it is normal for the action itself to constitute its own single point of reference for the constitution of equivalence classes, or he proceeds by anticipatory examination and then generalizes what he has found. In both of these cases, the subject attaches more and more importance to the specific nature of the elements and consequently to their objective similarities and differences.

In the first stage, the subject proceeds by trial and error until he has finished covering the bases which certainly presupposes a function of correspondence or of 'application' of the sets $X_{1,2,3}$ onto sets $Y = A, B, C, D$. But it is only as a result of the action of covering the bases that he discovers that a given element X_n can only correspond to one base, which he does not realize when he is only comparing the figures (anticipatory inspection), and he does not use what he has already observed in subsequent attempts. In other words, the subject still has not perceived the generic role of G in the bases (Figure 4) nor its correspondence with the V segment of the elements:

JEA (5; 6) starts out with B_2 and ends up by placing it on B: *'It hides the white'*. He proceeds by analogous trial and error for A, then for C_3 and says: *'It's O.K. – How did you see it? – I saw that the little feet* (extremities of H) *could go there. – Why? – The big piece hid the white* (V and G)'. But this discovery after the fact does not guide him in choosing a 'suitable' element for D, which he again finds by trial and error; neither does it help him to understand the one-to-many correspondence between the bases and the remaining elements (surjection): 'What about the others (there are only 6 left of the 10 he was given, excluding the identical ones)? – *There are no more.* – Do you think there is only one piece for each? – ... (he tries A_2 and finds it): *It works.* – (He continues by trial and error and finds C for C_3) – Do you think it won't go on another? – *We'll see* (tries on B): *No, it doesn't work.* – Why? – *Because it's turned around'*. He points to M which is turned towards the left, without seeing that this is of no importance and that it is only the position of V in relation to H which matters.

CLA (5; 7) also proceeds by trial and error and to indicate the correspondence between C_3 and C only finds that: 'There (H) *is flat and there* (v) *goes up just a little'*, although it is in fact V and its relative position to H which matters 'and that one (C_2)

can you say where it goes without trying it (thus by simple perceptual inspection)? – *No* (he tries on *B* and says): *that should come out farther* (= *V* more to the left). – And that (C_3), can it go any place else? – *It can go on* (*B*). – Try it. – *No, it won't work*.

FRA (6; 2) proceeds likewise and as a reason he simply says: '*I look up there* (he shows the base in *H*), *that's how I know*' but each time fails in his anticipation and says: '*I have to try it*'.

As we can see these subjects only arrive at the correspondence or application by relying both on the action itself and on its results, without anticipation by perceptual inspection, without generalization for subsequent cases and furthermore without finding the reasons for the 'suitability' once it has been discovered through the action. The subjects of the following stage, on the other hand, are able to isolate some of the reasons for their empirical success although they do not anticipate nor generalize the results in subsequent attempts:

CAT (6; 2) begins by displaying some stage I reactions, but, after some success by trial and error, she says: '*I look to see if it's that way* (she shows *H*). – Does it hide the white (suggestive question intended to help her)? – *No, it's above all that* (*V*): *you have to look to see if it's there* (position) *and if it works* (*G* on the base)'. She has thus retroactively grasped some of the reasons but fails to apply this discovery to the next case: 'Does C_2 go somewhere else? – *Maybe there* (*B*: she tries). *No, it doesn't go there*'. The role of the position of *V* has thus not been understood.

DOM (7; 9) does not anticipate anything and proceeds by trial and error, but quickly discovers that '*to put* (C_1) *on* (*C*), *I must put* (*V*) *here* (shows the place of *G*). – What does it cover? – *That* (*G* on base *C*). – And can (C_1) go anywhere else? – *Yes, on* (*B*: he fails). – Only on (*C*)? – *Yes*, – And that (B_1)? – *It also goes on* (*C*)'.

CRU (8; 2) after repeated trial and error, finds that '*the big part* (*V*) *hides this* (*G*)', but immediately thereafter foresees that C_2 will go on *D*, etc.

OLI (9; 4) after trial and error says: '*I look at the space to see if the red one* (*V*) *and the white part* (of the base) *are in the same place*'. This would seem to indicate the position of *V* applied on *G* but it in fact only refers to the empty intervals found to the left and right of *V* and *G*. He does however predict that B_3 suits C_1 and C_3 suits *B*, etc., but each time has to try them in order to check for incorrect anticipations.

In short, the progress evidenced in this second stage consists in the discovery of the correspondence of *V* and *G*, in that *V* can cover *G* independently of *M* and *v*. On the other hand, the role of the position of *V* or *G* in relation to *H* or to the edges of the large square framework has not yet been perceived, in spite of the overall evaluation of the empty 'spaces' (Oli). It is not until the third stage that this necessary and sufficient condition of a common position for *V* and *G* in relation to *H* is discovered and immediately generalized. As regards the initial period of trial and error, it may be longer or shorter or even reduced to a *minimum*, with comprehension and anticipation occurring almost from the start:

REN (8; 9) after lengthy trial and error: *'There's the bar (V)!* – Did you find the trick right away? – *No, I discovered it afterwards'.* He then succeeds without error with C_1, B_1, D_2, B_3 and A_1: 'How many shapes are there per card? – *Three'.*

CHA (8; 10). Less initial trial and error, then: *'There (G on C) it's on the left, there (B) it's on the right. The white is not hidden on that one (B with C_3), it only goes on one'.* The rest are correctly anticipated and generalized.

NIC (9; 9). Little trial and error: *'I look at the large vertical bar (V) because the little white square (G) is there. The big red line (V) can only go on a single card, since the little white square (G) is moved (from A to D)'.*

JAC (9; 10): *'The space to the left and to the right is the same* (between an element and its base). *This little square (G) belongs in another place'.* Using the same formula as Oli in stage II, Jac thus finds the role of the 'place', and then goes on to make correct anticipations, even for the polyvalent elements E_1 and E_2 (as did the preceding subjects).

These facts regarding stages I–III already show us how the subject passes from a given action whose scheme constitutes a function, to detailed surjective correspondences. Here the action consists in 'applying' one object (figure $X_{1,2,3}$ ...) onto another (figures $Y = A \dots D$). It could also consist of imitating the object (thus applying it onto itself), finding it (applying it onto itself in another position), transforming it (applying its state a onto a state b in accordance with certain modifying or conserving dependences), etc. In this particular case we are dealing with the simple application of one surface onto another (the conditions being defined by colors) and the following will show how the subject passes from an action to a correspondence.

At first, the correspondence is not distinct from the action of application and both remain global (application of the whole onto the whole). Only the results of the action show (after the fact) whether or not the conditions have been satisfied. Thus trial and error continues until success is achieved but without the modification of the undifferentiated nature of the application. Correspondences are thus established step-by-step by pairs, with all of the pairs being similar since all are the result of the application of the same scheme of action ('applying X onto Y'). The objects intervene only at the end of the action to divide these correspondences into incorrect (provisionally, prior to correction) and correct ones. The justification in turn invokes only the 'irrelevant' properties of these objects (M, v, etc.). In stage II, on the other hand, the application, while conserving these general characteristics, is enriched by an object-to-object correspondence on a differentiated point: e.g., the application of V onto G. We therefore have a semi-undifferentiated application, i.e. the application of a scheme of action, and a semi-differentiated one, i.e. the term-by-term correspondence of the V and G

elements. This is how (still after the fact) the pairs (A, A_n), (B, B_n) are constituted, but for lack of generalization of the formatory pair (V, G), these pairs remain juxtaposed bijections as when $n = 1$ in X_n. In the third stage, there emerges a new differentiation between the positions of V and G in relation to H, thus making possible the generalization, anticipation and constitution of co-univocal correspondences or surjections: to 1 $Y (A \ldots D)$ there correspond 3 X's $(A_{1,2,3} \ldots)$.

3. THE SUCCESSIVE FORMS OF EQUIVALENCE CLASSES

This development of functions, through the passage from a global pair repeated n times, to differentiated but uncoordinated pairs (stage II) and to pairs coordinated by a surjection, is then translated by the constitution of equivalence classes of three successive forms.

In stage I, the subject starts out with the hypothesis of a total undifferentiated class determined by the scheme of the applicatory action when he first attempts to cover any $Y (A \ldots D)$ by any $X (A_{1-3}$ etc.). However, by trial and error he discovers the existence of 4 classes corresponding to 4 bases, such that the sole adequate definition of these classes is still that of an assimilation to the scheme of the action, but now a successful action. The most salient indicator of this centration on the action without as yet sufficient decentration on the object, thus of these 'suitabilities' relative to any action (see the Introduction to this chapter) in contrast to objective correspondences, is the fact that the subject, when asked if two X's are 'alike' (question II: §1), answers quite naturally in the affirmative since all he has to do to make them so is to modify one in such a way that there will no longer be any difference:

JEA (5; 6) already cited in §2, for example, compares C_1, C_2, and C_3 which we present asking: 'Are they a little similar? – *Yes, because if they were longer, these feet* (the right extremity of H which seems longer in C_1 since M is placed above it) *would be the same*. – But are they somewhat similar (we try in vain to make him mention V)? – *If we cut them, they would be the same size*'. Similarly comparing D_3 to D_2: 'Can we say that they are of the same family? – *Yes, more or less.* – Are they cousins? – *Yes.* – Why? – *If we cut it* (extremity of M) *they would be the same*'.

MAR (5; 7) for D_1: '*That (M) is too much, it shouldn't be there*' and the elements B_{1-3}: '*it (V lower in B_2 than in B_1 and B_3) should be higher and (M) smaller.* – But do they look a little alike? – *Just a little.* – Why? – *If we put that (M) up there* (top of V), *it would be the same.* – And this (A_1, B_1, C_1)? – *We would have to move it over here* (V in C_1 on the right side)'.

PHI (6; 2) for B_2 compared to the other B elements: '*Yes, that (V) should go up to the top*'.

FRA (6; 2 already cited in stage 1, §2): *'If we changed the square (M), it would work'* and for C_3 (Figure 2): *'If we turned (M) around, it would be the same as that (C_2)'.*

We are all familiar with the story of the little boy who, on seeing a small grey cat, insisted that it was a big brown dog 'because I can make him larger, cut his whiskers and paint him brown'. Our subjects tend to proceed in the same manner. In a more serious vein, however, their reactions call to mind those of children aged 4–5 who, in experiments involving flotation, hold a board underwater for a moment so that it will stay under and then use that to verify their previous prediction.[3] In other words, since the inherent nature of actions is to modify their objects, the action of application is commonly conceived in this way, and since the initial equivalence classes unite those elements which 'suit' the original base without yet taking into account objective similarities or differences (cf. Introduction to this chapter), the passage from 'suitability' to similarity is thus first assured by the simple mental modification of the objects which in turn permits their assimilation.

In the second stage, the appearance of objective correspondences leads to the search for common qualities, thereby freeing equivalence classes from the general scheme of actions as such, in favor of the establishment of a direct relationship between the objects. In other words, given the fact that the application of the X elements onto the Y bases is now beginning to take into account their differentiated characteristics, this application is generalized to possible coapplications of certain X elements in preference to others, thus resulting in the determination of their similarities. But since the X elements, i.e. $A_{1...3}$, $B_{1...3}$, etc. are still being manipulated and thus remain proximate to the action while the Ys $(A ... D)$ only constitute points of applications, we still do not have a sufficient comparison among the Y bases themselves in stage II. The fact that all have a G in common is definitely noted but the different positions of the G elements are not. In summary, the classificatory reactions of stage II lead to equivalence classes based on the similarity of the Xs independently of the characteristics of the Y bases (likewise between the elements $A_{1...3}$ and $B_{1...3}$, etc.). It is possible for one of the 'suitabilities' common to certain Xs in relation to one of the Y bases to intervene as one of the possible similarities, but only as one factor among others:

DOL (7; 4) in his comparisons, first says: *'Yes, it's the same ($B_{1...3}$, etc.): there are two large surfaces everywhere (H and V) and a small square'*; or for C_2 and B_1: *'They both have strips and a square (M)'*. But later he argues: *'They are similar because all three go on the same card'*.

NAT (8; 3) also says (for B_1 and C_3): '*That and that are the most similar because of their shape*' (and for $D_{1,2,3}$) '*They are a little bit similar because all 3 go on the same card*'.

But the equivalences thus constructed lead only to more or less juxtaposed classes having almost no hierarchical structure, which is natural since the characteristic common to the X's belonging to a given Y base, i.e. the constant position of V relative to H, has not yet been discovered. The quality of belonging to a given base, as invoked by Dol and Nat, is therefore only the empirical manifestation of kinship. However the principle involved still continues to elude these subjects and control is only possible by global perceptual inspection and by reference to somewhat vague characteristics such as the 'same shape' or 'both strips (V and H) and a square (M)'. These characteristics can also be found in the comparisons between elements corresponding to different bases (e.g. B and C), where the classification remains at the level of a juxtaposition of classes or collections, which from the standpoint of the object is comparable to the juxtaposition of initial 'suitable' pairs established from the standpoint of the action (cf. the correspondences of stage I, in §2). In short, this stage is only an intermediate step between constitutive functions of application and constituted or operatory functions, due to the absence of hierarchical class inclusions.

In stage III, we arrive at a hierarchical classification, i.e. at 4 classes which correspond to bases $A \dots D$, characterized by the position of their element G. These classes are disjoint except with regard to the polyvalent elements E and when they are united in a total class, as defined by the correspondence of the G and V elements. But we must clearly understand that in this case there no longer exists only a correspondence of 'many to one' between the $A_{1,2,3}$ and the base A, etc., but also a 'one to many' correspondence for each G, distinguished by its unique position relative to several V elements corresponding exclusively to that G. It is by virtue of this double reversible movement 'many to one' and 'one to many' that the inclusion of the elements in the 4 classes A, B, C and D takes place, and in turn the inclusion of the latter into a single class characterized by the law governing the position of the G and V elements becomes possible. This does not however constitute a return to the situation of stage I or to equivalences subordinated to the schemes of the action: If $A_{1,2 \text{ and } 3}$ are to be classified under A, it is due to the law involving the progressive displacement of V in relation to H and G, thus due to the properties of the objects as such. Not only do the subjects refer explicitly to these characteristics of position, but at times, in order to verify their common characteristic, they even proceed by superimposing them, whether by placing the elements on the same base as a means of control,[4] or even by applying them one on top of the other:

GRA (8; 9) unites $B_{1...3}$ which he distinguishes from $C_{1...3}$: 'Can we see right away if they are similar or must we try it? – *We can see.* – How? – *With the ends* (the V's). – What about them? – *In (C) they are closer together, in (B) they are more to the left.* – What is most important? – *The ends (V).* – And that (M)? – *That doesn't do any good'.*
          COR (9; 5) compares $B_{1...3}$: *'They go well together.* – Why? – *They have the same bar (V).* – How can you tell? – *We can put them on top of there* (shows base B)'. After having established the four main classes, Cor shows the analogy between certain A and B elements and certain C elements which also 'go well together' but from the standpoint of M or V. As regards the E elements we can put them '*with many piles'.*
          DAV (10; 11) compares $D_{1...3}$: *'They all have a strip in the same place (V* in relation to H). – Are you sure? – *Because if one is put on top of the other, (V) is in the same place'.*

These stages show the passage from functional schemes to operatory groupings, through the incorporation of one-way applications into a reversible operation whose correspondence is co-univocal in both directions: 'many to one' (application), and 'one to many' from which we obtain the additive inclusions of extensional classes.

          Constitutive or preoperatory functions thus become in stage II, and more so in III, constituted functions as a result of their introduction into a two-way operatory system.

4. COMPLEMENTS

Two questions arise as regards the white complementary elements X', i.e. their relation to the red elements X and their internal equivalence relations.

          In stage I the subjects succeed by trial and error in filling in the intervals between the partial surfaces of $A_{1...3}$, etc. with the $A'_{1...3}$, $B'_{1...3}$, etc., but without understanding the general relationship of complementarity nor seeing the properties common to the complements:

JEA (5; 7) sums up the task quite well when he asks: *'Those which close the transparency?'* Yet, once he has successfully completed the assigned action, he finds nothing in common between the complements of a given base, except (as in §3 where we saw his reactions) when mentally modifying the shapes: 'Is there something similar between those (the A' elements)? – *If we changed the white ones a little they would be the same'.*
          CAT (6; 2 who started with stage I and passed to stage II for the red elements) also succeeds in the task through empirical attempts: 'Can we put these red ones in the same pile $(C_{1,2,3})$? – *Yes.* – And the white ones $(C'_{1,2,3})$? – *No, because they don't look alike at all'.*

In stage II, the subject still needs to proceed by trial and error in order to put the white complements X' in the spaces free of red X elements, due to his inability (as we saw in §3) to realize how the position of the bar V relates to the horizontal bar H. He does however begin to understand the relationship of complementarity and bases himself upon it to isolate one of the common elements of the complements X', i.e. the presence of a bar V', although he does not yet take into account its position in relation to G (or to H):

DOL (7; 4 already cited in §3) succeeds in all his attempts: 'Could we place (D'_2) with those (two B_n)? – *No, because here* (B_n) *the little end* (*M*) *is on the same side*', etc. – We start again with the set of C_{1-3}. – 'Are the red ones similar? – *Yes.* – And the white ones (*C'*)? – *They're not similar* (remember that complements comprise discontinuous elements). – Can't we put them together? – *Yes, because there is a transparent bar* (= part V which is red on the elements $C_{1,2,3}$ but which remains empty, covered by transparent plastic, separating the white parts from the complements $C'_{1,2,3}$. – What does the transparent part represent? – *What we had in red* (on $C_{1,2,3}$). – Is that what makes them look alike? – *Yes, there are two transparent bars in all of them* (*V* and *H*). – And anything else? – *Also a big end and a little end* (the white surfaces of $C_{1,2,3}$)'.
    SYL (7; 10) likewise succeeds and indicates how she proceeded: *I looked to see if there was a big hole on* (B_3)', or '*that* (the white surface on *C'_2*) *must be transparent*'. As for the similarities, they can be put together '*because they all have a bar* (she shows that V is transparent). – Can we put the reds and whites together? – *Yes*'.

It is interesting that this relationship of complementarity begins to be understood around age 7–8, which is the same level at which it intervenes in the groupings of classes composed of discreet (spatially discontinuous) elements. But it is only with stage III subjects, who can construct hierarchical inclusions, that complements form an explicit subclass with common properties which correspond to the subclass of red elements ($A_{1,2,3}$ for A', etc.), both being subsumed under the general class defined by their common base (A, B, C, or D). Here are some examples:

CHR (8; 10) finds that the complements '*make the same type of design, because the red ones* (*X*) *go with the white ones* (*X'*)'. The classification thereby obtained comprises the subclasses X and X' for each base.

    COR (9; 5 cited above in §3): '*They don't look alike* (perceptually) *but when we put them together, they do look alike.* – Why? – *If there weren't all those holes, they wouldn't go together.* – They go together only when we do what? I don't understand. – *Only when we place a red one with a white one. The reds are not the same* (as the whites) *but with the whites, you get the card* (= the total coverage of the base)'.
    RIN (9; 11): '*They have to look alike since those* (*C_2* and *C_3*) *are alike.* – What makes them alike? – *The spaces* (he shows the empty spaces in C_{2-3} and the white ones in C'_{2-3})'.

NOV (10; 7): They are similar *'because the white frames* (= closes) *the transparent red. If we put a red one with a white one, they form a group. The surface which is not covered on the red corresponds to the white surface'*.

As we can see, the reactions to the complements confirm what we saw in the classification of the red elements (X). As regards the relations between the classification thus achieved (inclusion of subclasses and of their complements into whole classes) and the functions constituted by application, it must be pointed out that when the X elements are determined by the direct application of these X elements onto the Y bases, the complements X' do not constitute an application F' on the Y bases but do so only on the empty parts of the X's.

Thus, the following correlation and synchronism are of interest. When the application F of X elements onto Y bases becomes both generalized and anticipated (by univocal correspondence between the V bars and G parts of the base as they relate to the position of H), the subclasses of the X' complements are coincluded with the X elements into general classes (which the subject Nov calls 'groups' in the sense of entire sets). In other words, the operatory grouping of classification is here born of the composition of two functions, the application F of the X elements onto the Y' elements and the application F' of the X elements onto the empty spaces of the X' elements, thus permitting the total application of the $X + X'$ elements onto the Y bases.

5.   CONCLUSIONS

Our problem was to understand how a subject passes from simple constitutive functions of application to equivalence classes which could be treated operatorily in hierarchical inclusions (the grouping of classifications). The stages observed provide all of the transitional terms needed to explain this passage.

The initial stage is the function $y = f(x)$, the simple application of x onto y. From the viewpoint of the subject at that point, the functional dependence linking y and x is expressed by any 'suitability': a hammer and a nail, etc. and, in this experiment, by covering the base cards Y = A, B, C *or* D with the red elements ($X = A_{1,2,3}$; $B_{1,2,3}$; etc.). In these cases, several different x elements may satisfy a given action: an axe may replace a hammer or, in our experiment, A_2 may substitute for A_1. Given such an action F (a specific example of the function f in $y = f(x)$), the equivalence of the x's then only relates to the use made of them (cf. the 'definitions by usage' cited in the Introduction) in the accomplishment of the particular action F. We can thus see that the equivalence is as yet only determined by the action (it being understood that the objects x are its vehicles). Psychologically, therefore, it consists in an assimilation to the scheme of the action F (the source of a function).

Stage II involves the passage from equivalences based on the nature of the action F ('same suitability') to equivalences based on the properties of objects (similarities between x's). Psychologically this passage is easily explained if the initial equivalence is an assimilation of the x's to the scheme F. All assimilations are in effect repetitions, generalizations and recognitions, such that when the assimilations of x to F are no longer achieved by contingencies but rather systematically, the problem becomes one of recognizing the x's, and the subject is led to isolate their common characteristics. But this systematic recognitive assimilation of the x's among themselves is not established all at once, and prior to becoming both generalizer and anticipator, it proceeds in a retroactive manner, still based on the successes or failures with regards to the action. Thus in stage II, the subject isolates a single common characteristic (existence of V elements) instead of the characteristics taken as a whole (position of V elements in relation to H and above all to G). This level thus corresponds to that of definitions by kind alone, without specific differences (a mother 'is a lady'). This kind is subdivided not by hierarchical inclusion but by the simple constitution of small equivalence classes by direct *individual unions* of elements, which mathematicians call 'partitions'.[5] On the plane of spontaneous classifications, this level would thus correspond to that of non-figural collections without as yet any inclusions comprising a quantification of extensional inclusion ($X < Y$ if $Y = X + X'$).

The passage from this intermediary stage to the operatory grouping of classification poses a problem since the equivalence classes formed by simple 'individual unions' (the most simple of these being pairs) do not divide themselves into whole classes or subclasses as long as the new operation, i.e. class union, does not intervene. The latter imposes itself (thus proving its presence in stage III) in the case where a special class like that of $A_{1,2 \text{ and } 3}$ is compared to its complement $A'_{1,2 \text{ and } 3}$, from which stems the characteristic class union operation of stage III: $X + X' = Y$. What is the source of this operation, not yet implied in one-way applications of the type 'many to one' which existed prior to this level? The discovery characterizing stage III is the distinction of the various positions of the V elements (in the X's) in relation to those of the G elements in the Y bases, which leads to the correspondence of several X's to a single Y thus permitting the partition of Y into four distinct terms of application. But this does not suffice for a system of included classes to be drawn from these applications, even though, as Grize points out, an application defines a quotient-set on the domain (X) and a subset on the range. In effect, in order to pass from these applications to a classification or grouping of classes in accordance with diverse inclusions, it is not enough to apply X's onto Y's in 'many to one' correspondences; it is also necessary to utilize the reverse process and to constitute 'one to many' correspondences (i.e. the principle of the subdivision of a genus into its species). In stage III, it is this principle which specifically permits the subject to start with each of the four forms A, B, C, D characterized by the position of their element G, make them correspond to their A_n's, B_n's, etc., and unite them into subclasses according to the position of their V elements. But this 'one to many' correspondence is no longer an application! On the contrary, operatorily, it is the reciprocal of the application 'many to one' and together with it characterizes the general operation of co-univocal correspondence on which the groupings of classification are based. In short, the passage from a constitutive function to an operation is a process which completes a one-way application when it is introduced into a reversible system, and this reversibility is the source of inclusions. It is in fact to this operation that we refer implicitly when we pass from applications to quotient-sets and to the set of the subsets, since both basically imply the relation of the whole to its subclasses, i.e. a correspondence of one to many.

NOTES

[1] With the collaboration of E. Schmid-Kitzikis.
[2] See Inhelder, B. and Piaget, J., *The Early Growth of Logic in the Child*, Routledge & Kegan Paul, London, 1964.
[3] They also attempt to use their hands to impose an inclined or horizontal orientation on the arms of a balance, etc.
[4] It should be noted that the bases remain on their side during the comparisons or classifications.
[5] But we have long used this term to designate something else: the infralogical partitions within a continuum, thus opposing the parts of an object to the subclasses of a set. Here we will therefore say 'individual unions' (cf. 'partitions' in the mathematical sense), distinguishing these individual unions of elements from the unions of classes or subclasses.

CHAPTER 3

FROM REGULARITIES TO PROPORTIONALITIES[1]

If functions express the links inherent in schemes of actions and thus constitute the common origin of operations and of causality, it should then be possible, starting from elementary functions of application, to conceive of two modes of composing functions which are distinct, although naturally comprising numerous relations among themselves. The subject can, on the one hand, compose functional dependences between objects and thus orient himself towards the comprehensive systems of physical determination which assure their causal explanation. On the other hand, by coordinating his actions among themselves, he may also arrive at operatory compositions whose inherent nature is precisely to express the most general coordinations of actions. However, before arriving at these major operatory structures which are both closed[2] and reversible, the subject can and must make all sorts of less general coordinations whose compositions remain variable and open and cannot as yet engender comprehensive systems comprising laws which are also systems.

The partial coordinations of actions, whose compositions are multiple and even undefined and by the same token not closed, can be expressed in terms of the 'combinators' proper to combinatory logic (combinatory in the current sense of the term, not in the sense of n to n combinations etc.): i.e., combinators of identity, repetition, substitution, association, etc. Such a list is open-ended for the above mentioned reason. We will therefore hypothesize that the subject is capable of constituting and even composing functions by coordinating actions thanks to these operators well before the operatory level. These operators then engender certain functional regularities which in turn are capable of subsequently serving as points of origin for operatory constructions. In the following, we will refer to these combinators as *coordinators* since that is precisely the psychological role they play at the level of observed behavior and since logical 'combinators' often require individual treatment at the level of formalization.

Regularities due to coordinations between functional schemes could take the following form, among others: if $y = f(x)$ and $y' = g(x')$, where g and f are in a certain regular relation (of which the identity $g = f$ is a specific case), there will then exist relations not only between x and y or x' and y' but there will also exist relations of relations or composed functions such as 'y' is to x'

as y is to x'. These relations of relations, which the psychologist Spearman terms 'correlates' (for example: a beak is to a crow as a snout is to a fox), are not as yet proportions — far from it — since there is no equivalence between the cross-products. We can, however, refer to them as 'preproportional'. Our second hypothesis will then be that these preproportionalities which result from compositions assured by combinators (thus by the coordinations of schemes) must sooner or later give rise to an operatory treatment which will result, in its most regular form, in the notion of proportion in the strict sense of the term. Grize has shown how often this notion has been linked historically to the elaboration of constituted functions. (This is so true that when we commonsensically say that a repeatable relation is 'mathematically exact', at least nine out of ten times we are thinking of a proportion.)

This chapter will be divided into three parts. The first will examine the regularities of a spontaneous action, i.e. the child is asked to line up balls of two colors without being allowed to see the parts already constructed (as each ball is placed into a hole, the contents of the previous holes are hidden). The regularities observed in this particular task are interesting in that they constitute themselves and are derived one from the other by means of 'coordinators' which are not yet operatory (in our sense of reversible structures), but which are increasingly complex and lead, for example, to sequences such as abb, abb, etc. The second experiment will start with the sequence $1\,a\,2\,b$, also placed in rows with the emphasis on the holes instead of on the objects. A series of objects a, b, c, d ... and their counterparts a', b', c' ... (alike in all respects) will be presented by the experimenter (the sequences are no longer spontaneous), such that the elements a, b, c ... will be placed one per hole and a', b', c' ... will be placed every other hole. The task will then be to anticipate the position of the object n' by seeing where the object of rank n is placed (the preceding holes being empty). We will thus see how correspondences which are at first simply ordinal ($n' = n + 1$) and then hyperordinal ($n' = kn$ where $1 < k < 2n$) result in proportionalities. But since the rules of this game are arbitrary, there will be a third experiment in which the subjects will be presented with three fish ('eels' of equal diameter but of varying lengths) which must be fed (pellets of the same thickness but varying in number, or biscuits of varying lengths) as a function of their size, thus giving rise once again to the problem of proportionality.

1. FIRST EXPERIMENT: SPONTANEOUS REGULARITIES

The apparatus used consists of a long box measuring 30 x 3 cm which is placed horizontally and is subdivided into 24 successive compartments called 'holes' into each of which the child must successfully place one ball chosen freely from a reserve containing 25 white ones and 25 red ones. As the child fills each 'hole', a sliding cover is drawn over the compartments which have already been filled thus hiding them. Once this box has been completely filled, the child is presented with a second box just like it and is asked to 'do the same thing again'. Next a verbalization of the task is suggested: 'What have you done? Explain how you did it'. The child is then asked several different times to do 'something else, something different', etc. Finally, there is a test involving possible transpositions, e.g., the replacement of the red and white balls by cards of the same color but of two different shapes (25 squares, 25 triangles), by modeling clay balls in two sizes (25 large ones and 25 small ones), etc.[3]

I. – The first results obtained by F. Orsini are that the spontaneously constructed sequences present remarkable regularities: of 30 subjects aged 3 to 8 in the 8 original series (thus excluding the transpositions), we observe regularities of 58% at age 3–4, 85% at age 5–7 and 90% at age 8.

Secondly, these regularities are almost exclusively of three types: (α) First there are uniformities, i.e. sequences of the same kind of objects (all white balls, etc.) which prevail over the types at age 3–4 but which may still be sporadically observed in one-third of the 5–8 year-olds; (β) Next come simple or homogeneous alternations (one red, one white or 2 red, 2 white, etc.). Already present in one-third of the subjects at age 3, these regularities increase with age until they are found in 28–39 out of 30 subjects aged 5–8; (γ) Thirdly, there are heterogeneous alternations (1 red, 2 white or the opposite, 2 red, 3 white, etc.) which are found in 1 or 2 subjects out of 30 at age 3–4, in 10 to 13 subjects out of 30 at age 5–7 and in 23 of the 8 year-olds.

The reproductions and verbalizations of these regularities are relatively good. The latter consists of simple enumeration in more than 70% of the cases at age 3, 50% at age 4 and only 14% at age 9, after which they give way to the formulation of the law followed. Lastly, as regards transpositions they are only successfully made by 2 to 7 of the subjects out of 30 at age 3–4 but by 19 to 29 of the subjects aged 5–8 out of 30.

Let us now try to isolate the types of actions which could intervene in such regularities and translate them into a language of *coordinators*, or elementary schemes, capable of constituting functions, either separately or by progressive composition. To this end, let us recall that these regularities follow one another in a rather constant order of formation (F. Orsini found this to be true even in the case of mentally retarded children taking into account their level of development) and that the type (γ) (heterogeneous alternation) appears only after the constitution of a family of simple (β) alternations: 2–2, 3–3, 4–4, etc. (generally in this order), which would seem to point to the existence of a compositional process linking one type to the next:

(1) The initial reaction is a random alignment which although preceding regularities naturally comprises partial ones (it would be interesting to analyze these since it is doubtful that pure chance exists in psychology). We will limit ourselves to the basic coordinator presupposed by this alignment which we shall call W = *repetition*. It consists in repeating only the action, in this case, placing a ball.

(2) Now, if the repetition W expresses the reproductive assimilation of the scheme of the action, there exists a second basic coordinator which expresses the recognitive assimilation and will this time focus on the object of the action, i.e., *identification* = I. This is how the subject recognizes the white balls a in random sequences from which is obtained the function of self-application $a = a$. This identifier I can also range from a pure identity (I_0) which expresses the action of finding the same object (permanence of a singular object), to the complete equivalence (I_1) bearing on objects which are equal but distinct (one a and another a) to equivalences of various orders ($I_{2...n}$) based on simple similarities.

(3) The α type regularity (uniformity aaa . . .) does not presuppose a new coordinator which would consist of 'adding the same', since it can already be conceived as the result of a coordination between W and I (here I_1), i.e. WI repeating the action by making it bear on the same element. And, since this repetition is continuous, we can write: $(WI)W$.

(4) Next, after having aligned the white balls aaa . . . , the subject is made to align another series of red balls: bbb . . . . Thus, we introduce a new coordinator[4] which we will call *substitution* = C. The series bbb . . . will thus correspond to $C[(WI)W]$.

(5) There then follows the β type regularity, or $abab$ . . . which as we can immediately see consists in combining aaa . . . and bbb . . . passing alternately from one element of the first sequence to an element of the second. It is thus a question of identifying a, or I, followed by a substitution of b for a, or CI, which is in turn repeated a given number of times, thus $[I(CI)W]W$, etc.

(6) From there the subject goes on to regularities of the type $aabbaabb$ . . . , or $aaabbb$, etc., which is the same thing as repeating a, thus WI, or repeating b, thus WC, within the preceding composition, thus $[(WI) \cdot (WCI)]W$.

(7) The γ type regularities (abb, etc.) are easily composed and we could also transform the set ab into ba by changing the order of the coordinators or of the actions. But it is advantageous to assign an inversion coordinator C_2, because the subject sometimes refers to it specifically and immediately: 'We must also do the opposite' (age 5), etc. From there the subject easily proceeds to symmetries such as $abba$ by reversing one of the pairs.

These 'coordinators' having thus been characterized, a function may then be defined as the link uniting the dependent elements (the physical objects or the 'terms' of a relation) on which such coordinators act within the context of compositions which are analogous to the preceding ones. This is exactly what we are saying when we describe a function as the result of an 'application', since, an application does not constitute a specific combinator but rather expresses the correspondence between the results of an action A_1 and those of another action A_2 (i.e. between the initial and final states of an object on which an action has been effected) when these actions have been effected or coordinated according to the combinators $W \ldots C$, etc. As a result we obtain functions of identity ($a = a$), of equivalence, bijection, etc.

II. − Let us also note that among the γ type regularities, there are some whose progression is additive, such as *abb*, *aabbb*, *aaabbbb*, etc.,[5] although some can also be multiplicative (e.g., *abb*, *aabbbb*) thus evoking proportionality. There is no reason for these to constitute long series, although it is easy to provoke, as it is to appeal to additive compositions in situations which suggest compensations. In complementary experiments which we will not here analyze, F. Orsini has, for example, used the preceding device to study two kinds of reactions: (1) If we give the subject (3 successive times) 1 (*a*) 2 (*b*), then 2 (*a*), how many *b*'s will he put down?[6] Does the same hold for 1 (*a*) 2 (*b*), then 4 (*a*) (or 3 (*a*), 5 (*a*), 6 (*a*) and 10 (*a*))? Or, if we place 1 (*a*) 3 (*b*) (3 times) then 2 (*a*) (or 4 (*a*), 3 (*a*), etc.), how many *b*'s will he put down? (2) If we place 4 (*a*) 1 (*b*) (3 times) then 3 (*a*) (or 1, or 2 or 0 (*a*)), how many *b*'s will he put down? We can see that the first group of questions suggests a proportion while the second suggests a compensation.

We will not address ourselves to the problem of compensations in this chapter for it will be covered in Chapters 5 and 6. Let us now merely point out that such simple compensations (giving *b*'s so as to conserve the equation n (*a*) + n' (*b*) = 5) become frequent starting from age 7−8.

As regards the suggestions of proportionality ($2\,n$ (*b*) for n (*a*) and 2 (*b*)), as a general rule, they only give rise to correct multiplicative reactions starting from age 10−11. Once the subjects have passed the stage of simple fluctuations or solutions by equality, the most frequent reaction is additive: 3 (*b*) for 2 (*a*) if 2 (*b*) for 1 (*a*), etc. But, in order to be able to analyze the constitution of these preproportions and their passage to the most precocious functional structures of proportionality, it would be advantageous to proceed by spatial correspondences or by direct correspondences between numbers and spaces, etc. It is for this reason that we will continue to use the apparatus in which the objects are placed or set in holes, but with variations in the number of spaces between the holes (§2). We will subsequently introduce cylindrical fish of 3 different lengths, numbered 1, 2 and 3 and ask the subjects to find the corresponding amount of food, thereby adding a causal dependence to spatio-numerical correspondences. This should result in the constitution of a proportional function as an extension of the correspondences themselves.

2. SECOND EXPERIMENT: THE INITIAL AWARENESS OF INCREASING DIFFERENCES

The child is given two strips each having 10 holes placed at an interval of 1 cm. He also receives a number of objects, which are grouped in similar pairs (A, A'), (B, B'), . . . and which fit on the strips when their stems are placed in the holes. The experimenter keeps one strip on which the objects A, B, C and D are placed, one per hole. The child is given the other strip on which the corresponding objects A', B', C' and D' are placed, one every other hole. We explain to the child that this is a game ('garnishing garlands') and that the object is to find the 'trick' in order to be able to continue. The child is made to reconstitute the sequence of elements $A'-D'$ until we are sure he understands it. We then start over placing A, C and D, then only C, etc. subsequently leaving both strips in which the holes 1, 2, 3, 4 and 2, 4, 6, 8 are filled in plain sight of the subject, so that he can refer to them. Once these preliminary explanations have been understood, the child is then given two strips I and II having 38 evenly-spaced holes and we proceed as before. Next we remove elements B, C and D, etc., leaving in place only A and A' (the 10-hole models are left alone), after which we place an object in the 6th or the 10th hole of I and ask the child to place the corresponding object on strip II in accordance with the rules of the game or the 'trick' (which means he should choose the 12th or 20th hole in II). Once the subject has given us an answer, we insert some objects between holes 1 and 10 and ask for the correspondences as a control. Next we place an object on the 19th hole (middle of I) asking for the correspondence in II (= 38th hole, i.e., the last one). We always request a justification and a repetition of the rule or 'trick'.

The 10-hole rulers are lined up with each other with a space between them. Those with 38 holes are partially superimposed (I to the left) but with an interval between the points of origin so as to avoid a direct visual correspondence.

Among 12 children aged 6, 10 aged 7, 10 aged 8, 10 aged 9, 13 aged 10, 11 aged 11 and 10 aged 12, we find three types of behavior. These correspond to two very distinct stages I and III, characterized by stable answers, and an intermediary stage which is less clearly defined and is often very transitory with some answers tending towards stage III or regressing back to stage I.

The first of these types of reactions consists of skipping only one hole. Thus for an element E placed on I in the Nth hole the corresponding object E' will be placed in the $N + 1$th hole of II. The second type consists of placing the object E' in the $N + k$th hole of II, k being a constant (or more or less constant, i.e. $k + m$) larger than 1 and smaller than $2 N$. Lastly, the answers of the third type are correct: if E is in the nth hole on I, E' will be placed in the 2 nth hole on II, by means of enumeration of the holes or by simple spatial duplication.

The distribution of these three types of responses is as follows (as a percentage with the number of subjects in parentheses):

	$n + 1$	$n + k$	$2n$
6–7 years (22 subjects)	91(20)	9(2)	0
8–9 years (20 subjects)	55(11)	15(3)	30(6)
10–11 years (24 subjects)	37(9)	9(2)	54(13)
12 years (10 subjects)	0	10(1)	90(9)

We thus see that the function $y = 2x$ is here only reached by half of the subjects at age 10–11 and by 9/10ths of those at age 12, while at age 6–7, 90% of the children invoke only the additive function $y = x + 1$. Although the intermediate response can be found at all ages, at age 6–7, it marks the limit of the child's comprehension while at age 8–11, it constitutes a mere step towards the discovery of an exact proportional relationship. The late emergence of this very simple spatial or numerical duplication at first seems quite surprising. We shall now attempt to find the reason for this by examining several cases.

The following are examples of stage I in which the function remains additive:

ANI (6; 10) understands the difference between the model strips: '*There* (the subject's), *some are missing* (holes without objects), *there none*'. Next we place an object in the 6th hole of I: Ani hesitates between several nearby holes (5th) then chooses the 7th. 'Why? – *We leave a space* (thus 6th empty + 1). – Must the space be bigger in yours than in mine? – *Yes*. – Much bigger? – *No*. – How much bigger? – *Two times bigger*. – Very good. Now is that right in that hole? – *Yes*. – Why? – (She shows the 6th on I)'.

CRI (7; 9): 'How should it be done? – *You should always skip one hole*', but he stops at $N + 1$.

GEO (8; 8): '*In yours* (I), *it's narrower, in mine* (II), *it's wider*'. We put E in the 10th hole, she puts E' in the 11th. We interpose new objects, which forces her to move E' back to the 13th hole. – 'So, what's the trick? – *It's more than one hole* ($N + k$)', but then she reverts back to $N + 1$.

RIA (8; 1): '*You always have to skip a hole*'. We put E in the fourth hole and Ria places E' in the 7th which would seem to be a duplication. 'How did you do it? – . . . – (We put E in the 19th). – (She places E' in the 21st). – (The experimenter moves E' on II to the 36th hole). – Is that right? – *Yes*. – Why? . . . – (E in the 9th). – (Ria places E' in the 10th and then stays with $N + 1$ from then on)'.

MAR (11; 3), who is an example at the other extreme of the ages studied, correctly says: '*You must always leave a hole between each object*' but for $E = 10$ she places $E' = 11$, etc. We interpose intermediaries, which forces her to push E' back to the 16th. 'Why?' – *Because you need a bigger gap*. – (E in the 19th: she places E' in the 20th and we push it back to the 38th). – Is that right? – *No, it's not: you need to have another hole*'.

These cases leave no doubt as to the child's comprehension of the model. Since the model is visible the subject sees that to positions 1, 2, 3, 4 there correspond the placements in the holes 2, 4, 6 and 8, which he expresses by saying that we must 'always' skip a hole (Ria), i.e. we must leave a hole 'between each object' and the preceding or following one (Mar), etc. Nevertheless, the solution found is only $y = x + 1$ and when the subject is momentarily diverted from it, whether it be by suggestion or because he finds the space to be insufficient, he reverts to it as the sole acceptable relation.

This type of reaction, already observed in many other analogous situations (we will again find it in §3), requires an explanation. It is in fact too easy to say that addition being a simpler operation than multiplication or the sequence of wholes being additive, the child is limited by economy or by simple confusion to using an additive composition, as if the former genetically preceded the latter. In reality, multiplicative compositions are inseparable from additive compositions with the former being no more than the repetition of the latter n times, since $n \times m = m + m + \ldots + m_n$. In this particular case, when the subject says explicitly that 'you have to skip a hole every time', this repetition is specially simple and it is what we will see stage *III* subjects do starting from age 8. Stage *I* subjects persist with the composition $y = x + 1$ for a reason which is genetically prior to the constitution of cardinal addition, i.e. because of the lack of differentiation between the latter and ordinal addition or the succession of ranks. The child simply reasons that if the holes in y are of a higher rank than those in x, all he needs to do is to add 1 to the final rank of x's, without worrying about what preceded it. This is the same error as that which we so frequently observed at this level in the evaluation of lengths, when the subjects judges a straight line segment y to be longer than another segment x (longer = farther) because it goes beyond it, without taking into account the points of origin. This error is due to the primacy of order over intervals (numerical or metric),[7] a primacy which is itself derived from the precedence of functions (= oriented applications or ordered pairs) over operations (which, because of their reversibility, lead back to the point of departure and require the subject to consider the intervals between the latter and the points of arrival). Thus, from the standpoint of order, there exists an ordinal addition, which generally corresponds to cardinal addition (except in the transfinite), but without any ordinal numerical multiplication. This results in the initial primacy of additive (ordinal) numerical compositions, while in the domain of operations of classes or relations, the multiplicative matrices are contemporaneous to additive class inclusions.

The next stage is marked by the solution $n + k$ which is similar to what we will see in Chapter 7 for the experiment involving the filling of jars in which the subject anticipates that the height intervals will be longer in jar B than in Jar A, but that all will be equal among themselves (case of Jac in §3). Once the subject has understood that there must be an empty hole between each element and the next, he comes to realize that instead of needing only a single empty space after E (solution $n + 1$), he will need several spaces. By the same token, however, he does not attain the function $2\,n$ because he fails to consider the transformation as leading to an *increasing* difference. In other words, contrary to stage I children, he admits that there is a transformation but he takes it all at once instead of seeing it as a cumulative process:

ELI (9; 11) begins with the solution $n + 1$: E' in the 7th for E in the 6th hole. But after interposing the first element, she says: *'That was wrong, I must skip one hole'*. Eli then moves the object back to the 10th hole: *'I must put it in the fourth one over'*. We place E in the 19th: she puts E' in the 23rd and so on. We place E' in II at twice the distance of E asking Eli what she thinks about this: *'I think that's right, because there is a greater distance* (with regard to $E = 19$). – What's the trick? – *You always have to skip 4 more'*.
    FRA (10; 1) for $E = 10$ places E' in 15. 'And if we put it here (11)? – *No, that would be too close.* – And here (20)? – *That would be alright; there can be several holes, but you need at least one'*.
    CLA (12; 3) for $E = 9$ places E' at 12. When we interpose several objects he moves it back to 16. 'How many holes are you leaving? – *Several'*. For $E = 19$, he rejects E' at 38: *'That's wrong, it's too far, I'd put it in the middle,* (he puts it at 30). – What's the trick? – *Each time you have to make it bigger, so that you don't have to move the objects* (if we interpose any between E and E')'.

Although the last subject has not yet attained proportionality, he is approaching stage III when he admits the existence of an increasing interval between E and E' when we move E farther away from the point of origin. Fra has almost reached this level as evidenced by his acceptance of an interval which is variable, but which does not yet increase regularly. Eli is typical of stage II in that she understands the fact that the interval is > 1 but gives it a constant value due to her inability to grasp the idea of a transformation which is variable even if not continuous. Thus it is the notion of variation of a variation which is grasped gradually through this intermediate stage.

The following are examples of stage III beginning with 8 year-old subjects whose age and mode of reasoning show that explicit multiplication is not essential for the solution of the problem, even though the functions used by the subjects gradually lead to the structure of proportionality:

AIR (8; 4): *'In yours, there are no holes but I have to have some in mine'*. We put $E = 6$ and she places $E' = 12$, etc., without trial and error. 'What's the trick? – *I don't know.* – If you had to explain it to a friend? – *She should skip a hole each time.* –

Explain. – *There are 6 holes in yours, for each one I need an extra hole in mine; there 6, so 12 in mine.* – How did you get that? – *We skipped a hole in mine*'.

ROL (8; 4). For $E = 5$, he places $E' = 8$: 'Is that right? – *No* (he places it in 10). – How do you do it? – *I count the holes. There are twice as many as in yours*'.

BER (9; 4). For $E = 10$, E' placed in 20: 'How do you know that? – *I counted the holes* (in I) *then I counted another 10* (in II) *because each time there is one between two*'. $E = 19$, E' placed in 38. 'Why? – *Each time we leave a hole between two, that makes an extra place each time*'. $E = 5$, E' placed in 10: '*Because there I counted 5 and here two times 5.* – How is that? – *Each time we add 1 between 2, that makes it double*'.

DOL (10; 2): '*If you skip to 7, I need 7 more, I must add. I must make this distance, then once again. We make twice the same distance*'.

ADA (11; 6): '*You must count how many holes were skipped, then count them twice*'.

ALI (12; 6) immediately sees a relation of single to double during the presentation and without hesitation relates it to *I* and *II*: this is the only case in which there is a direct intuition of the multiplicative relationship without first passing through the addition of the empty holes.

These cases, especially the first ones, are interesting in that almost all use the same formulas as in stage I: 'we skip a hole each time', etc. But from there they conclude the necessity of finding the 'double', and even if they construct it additively (cf. Ber and Dol), this constitutes a new element in comparison to stages I and II. What does it consist of?

In comparing the responses of the three stages, especially the intermediate reactions of stage II, we see that for the subject the problem is essentially one of establishing a correspondence between rows *I* and *II* while taking into account their differences. Initially, the correspondences and the search for an isomorphism are emphasized by minimizing the difference. In this case the minimized difference is first of all situated at the terminal limit of row *I*, which is also the end point of the action or of the oriented application which constructed it. Thus, reduced to its *minimum* the difference would be $n + 1$, in other words, only the last empty space would be considered. This produces the systematic responses of stage I in which the transformation of I into II is attempted by the simple addition of an empty unit. But since this initial function is the same as merely identifying rows *I* and *II*, with the sole exception of the terminal position, the subject, specially under the effect of the interposed elements introduced to make the problem explicit, arrives at the idea of a correspondence with transformation. However given the general attitude of simple 'application' of *I* onto *II* induced by the scheme of assimilation, the accommodation of this scheme, in the presence of this new transformational exigency, can only be carried out in a block-like manner, as a static difference which is given once and for all and leads to a still additive solution consisting of adding first a constant number of holes and then a variable number (non-systematically and by trial and error) but without continuous transformation. Stage II responses are thus still unstable, for once it has been admitted that the difference is greater than 1 and is subject to variation, the idea of a progression in the variation — thus a continuous transformation in relation to the number of holes characterizing the rank of E in I — gradually imposes itself. It is then, but only then, in constructing row II starting from I, that the subject sees the difference as increasing. He no longer starts with the terminal element E, but follows the order $A, B, C \ldots$ and thus takes into account the entire succession of empty holes, no|longer focusing exclusively on the last ones. Even if the procedure remains additive, it becomes cumulative at this point and ends in the function $y = 2x$. The subject finally becomes aware of this function when he uses the words 'to double'. This seems to be the history of this function, constituted starting from a simple constitutive function of assimilation or application of an initially global scheme.

It should be pointed out that this evolution seems dominated by a general process which can be characterized as the progress of reversibility on the sole initial proactive orientation. In stage I, in fact, under the influence of the essentially oriented functional 'applications' (= 'univocal to the right' according to their formal definition), the subject reproduces in II or y what he observes in I or x, but he makes the application in the direction of the order and focuses on the terminal limit, which results in $y = x + 1$. Afterwards, he goes back (stage II) to take into account the intervals and not only the order of the successive applications. Equilibrium is reached with the emergence of operatory reversibility when the reconstitution of the series (retroactive process) returns to the point of origin in order to take into account all of the intervals in such a way that they can be taken into account in the proactive application. Thus it is the subordination of constitutive functions to operatory reversibility which transforms them into constituted functions of proportionality.

3. THIRD EXPERIMENT: FROM SERIAL CORRESPONDENCES TO PROPORTIONALITIES[8]

The preceding structure already constitutes in certain respects a serial correspondence in that to the increasing distances between the element E and the point of origin on row I there correspond increasing distances whose increases between the element E' and the point of origin of row II become greater. At the beginning, the subject tends to merely transpose the same increase in distance from row I to row II and is only gradually able to establish a correspondence between the regular variation in I and the differential progression in II. As we can see, this covariation tends towards proportionality. In the first experiment, the only variable was distance, which could be ordered serially, while the objects $A, B \ldots E$ manifested no law of increase. We might therefore find it interesting to next examine a situation in which the objects given vary among themselves and can be placed in a series of increasing order on row I. The objects to be constructed on row II thus would also increase in size as well as proportionally in relation to the objects of row I.

The objects given on I are three fish 5, 10 and 15 cm long. So that the subjects will only have one dimension to consider, the fish are in the shape of eels with a uniform diameter. We do not insist on the exact measurement of these dimensions, but specify that given their differences in size, fish B will eat twice what fish A does and fish C three times that amount. This is tantamount to providing the solution, albeit only verbally. It is then up to the subject to determine on row B the amount of food, which is of two different types. In the first test, pellets are represented by 50 equal beads which are made available to the child. The task is to choose the appropriate number for each fish. In the second test, the food consists of 'biscuits' represented by strips of varying lengths, the problem here again being to make the quantities correspond to the size, i.e. the appetite of the fish, with the objects this time being continuous.

Let us again note that the differences between such a serial correspondence and the customary operatory tests used on analogous structures is that, in this particular case (as in that of §2), the elements of row II are to be constructed (they were intervals in the preceding experiment and here they are the objects themselves) not just ordered. When in another test we gave the subject 10 dolls, 10 canes and 10 backpacks of increasing sizes and asked them, for example, to find the pack and the cane belonging to the 6th doll, it was only necessary to seriate the three sets so as to make them correspond according to size (the same applies to double seriation matrices of two dimensions, such as size and color).[9] On the other hand since, in this case, the objects A, B and C on row I are already given, the subject must construct the corresponding objects A', B' and C' (number of beads or length of lines), which is in fact a functional problem: if $y = f(x)$ and x is given, the subject must determine not only y, but also the form of the function f, i.e. whether it is additive or multiplicative.

Of the questions asked, the six which are of interest to us are as follows (there were three other questions which involved dividing a cake but they are of no concern to us here and since they were asked at the end of the question period they did not influence the answers to the questions which preceded them): (1) If there is one pellet (= one bead) in front of fish A, how many are needed for B ('who eats twice as much as the little one') and for C ('who eats three . . .', etc.)? (2) If fish B receives 4 pellets, how many are needed for A and C? (3) If fish C were to receive 9 this time, what should A and B get? (4) Same as question (1) but in terms of 'biscuits'[10] with a strip 1 unit in length being given to A; (5) Same as question (2) with fish B being given the strip 4 units in length; (6) Same as question (3) with fish C being given the strip 9 units in length.

We were able to distinguish the four following stages. During stage I, the subject thinks only in terms of 'more' and 'less' and accepts all of the solutions, but only within certain limits and provided that fish B has more than A and C has more than B. (This holds at age 5–6 for the beads and strips and continues to 6–7 for the latter.) In stage II the subjects set up a numerical sequence of whole numbers 1, 2, 3 or 3, 4, 5 etc. such that the difference $\alpha = +1$ (beginning in certain cases at age 5–6 and continuing up to age 8 for the beads and even age 9 for the strips). For the subjects in stage III the difference is equal to $K \pm m$ where $K > 1$. The fourth stage can be subdivided into two substages such that in IV A one of the two relations (AB or BC) is correct though the other is not while in IV B both relations are finally understood.

As regards the first stage, let us first note that of the subjects tested between ages 5 and $5\frac{1}{2}$, three out of five duplicated the tasks correctly and four out of five knew how to count (one girl said: 'I know how to count to 20 and maybe more, but I have never tried to go farther'!):

CHO (5;6). Question 1: he gives 2 to B and 4 to C. 'Explain. – *It seems right.* – And if I gave 5 to C, would that be all right. – *Maybe, but not* 6'. Question 2 (4 to B): he gives 2 to A and 5 to C. 'Explain. – *Well, 2 for (A) since it is smaller and 5 to (C): it's bigger.* – Could you give 6 to C? – *Oh, yes!* – And 3 to A? – *No, that seems too much for the little one'*. Question 3 (9 to C): '*3 to A, then that* [5] *for B* (without counting). – If we give it 6? – *Yes, that's O.K. too.* – And 8? – *Maybe that too.* – And 9? – *No, that's not O.K., it would be the same as for (C)'*. Question 4 (strip 1 to A): he gives 4 to B and 6 to C. 'Why? – *I don't know'*. Question 5 (strip 4 to B): '*I give* [2] *to (A) and* [9] *to (C). But I wouldn't give that* [4] *to (B), it's the same as it had for breakfast* (question 4). *We must give him more.* – How much? – *That* [8] *for example'*.

KAR (5; 5 the one who knew how to count to 20 'or maybe more'). Question 1 (1 to A): She gives 2 to B and 3 to C '*because (B) is the medium fish and (C) is the biggest one'*. Question 2 (4 to B): '*2 to (A) and 5 to (C). – And if I give 1 to A? – No, it's not enough, it would be like the first time. – And 6 to (C)? – Yes that's good too. – And 7? – Also. – And 9 (to C)? – No, that's sure to be too much'*. Question 3 (9 to C): she gives 3 to A and 4 to B: '*If I give 6 to (B), would that be O.K.? – Yes. – And 6? – Yes. – And 9? – No, the big one (C) must have more: you said so*'. Question 4 (1 to *A*): Kar gives 2 or 3 to *B* and 4, 5 or 6 to *C*. Question 5 (4 to *B*): she puts down 1 for *A* and 7 for *C*. Question 6 (9 to *C*): Kar gives 7 to *B* and 1 to *A*: 'Is that enough for (*A*)? – *Yes, he's the tiny one*'.

We thus see that the principle underlying the dependences between the size of the fish and the amount of food required is that of a simple qualitative serial correspondence without the equalization of differences as we shall find in stage II where each amount (pellets or strips) will differ from the next by exactly one unit. In fact the lengths of the fish are too distinct (5, 10 and 15 cm) to impose the perception of an equal difference and the subject only has to concern himself with respecting the inequalities $A < B < C$ and to reproducing them in the relation between the amounts of foods. Likewise the child accepts almost all of the variations as long as no two of the fish have the same number of pellets or biscuits of the same length and the order $A < B < C$ is maintained. As regards the numbers, they do not yet have an iterative cardinal value ($2 = 1 + 1$, etc.) and are thus only ordinal symbols corresponding to qualitative sizes, so to speak (that's why Kar believes that she could maybe count beyond 20 if she tried!): thus the lack of numerical quantification of the differences. It is also true that these subjects sometimes refuse to accept just any size: Cho wants to give 5 pellets to *C* when *A* only has one 'but not 6' and if *B* has 4, the number 3 'seems to be too much for the little one'. The same applies to Kar, if *B* has 4, then 1 'is not enough' for *A*, and 9 for *C* 'is surely too much'. But these kinds of qualitative proportions are still only one kind of serial correspondence, i.e. the order $A < B < C$ must be maintained so that none has too little nor too much.

What then is the principle underlying these limitations, if the evaluations are not yet numerical but rather remain intensional or qualitative? Here, in contrast to the preceding experiment (§2) in which functions are derived from an arbitrary rule and in which number was somehow imposed by the device with its intervals and holes, the functional dependence linking the fish to its food is found, as in all elementary constitutive functions, midway between causality and an operatory relationship. From the standpoint of the causal context of feeding (which for the child is reinforced by social custom, thus the allusions to the difference between breakfast and the 'other meals' made during the task), we could just as well give $n + 1$ or $n - 1$ instead of n pellets to one or another of the fish according to its size, but too much is too much and not enough is not enough. Outside of these predictable limitations, qualitative estimations remain entirely unconstrained as long as they respect the order $A < B < C$.

Numerical quantification begins with the second stage, quite naturally in the case of the pellets or beads which are discontinuous and equivalent units while on the other hand the strips which are of continuous lengths resist an arithmetizable metric for a longer period of time. This numerical quantification begins in its most elementary form: the simple succession n, $n + 1$ and $n + 2$. Here are some examples:

JAN (5; 11). Question 1 (1 to A): Jan gives 2 to B and 3 to C. Question 2 (4 to B): '*One less for (A) makes three and one more for (C) makes 5*'. Question 3 (9 to C): '*One less for (B) is 8 and another less for (A) makes 7*'. In questions 4–6, Jan likewise predicts the strips 1, 2, 3 then 3, 4, 5 and finally 7, 8, 9.

   DEN (6; 1). Same reactions. For question 2 where he gives 3 to A and 5 to C when B has 4, we make him repeat the task saying B eats twice what A eats and C three times that, but he sticks to his numbers. 'Is that fair? – *Yes, 3 to (A) because it is a little fish and 5 to (C) because it is a big one*'. We then ask Den another question: if A has 3 and B has 6, how many for C? '*8 for (C). The first receives 3 because it is little, (B) has 6: double the amount. (B) has more than (A) and 8 is even more.* – And if you give 12 to (C) and 8 to (B)? – *Then 6 to (A). You skip two numbers, 9 and 10. Between (B) and (A) you skip one*'.

   With the strips, Den whom we tried to get to 'skip numbers' reverts to the system of adding units: if we give 2 to B, A will have 1 and C will have 3, if we give 4 to A, B will have 5 and C will have 6, etc. 'Is it really fair to give them 4, 5 and 6? Won't there be one who is hungry and another who has too much? – *Yes.* – Then what if we give 3 to A and 6 to B? – (C) *will have 7*'.

These reactions clearly correspond to those of stage I of §2. Yet while in the last experiment it was only a matter of continuing a given series which the subject conceived in terms of +1, the problem here is for the child to quantify the qualitative serial correspondence which he intuited at the outset (stage I of Cho and Kar, etc.). In this case this sort of qualitative or ordinal proportion which he perceives between the fish A, B, C and the order of the amounts of food A', B', and C' (where B' is to B as A' is to A, etc.) seems to him to be expressible in the simplest cardinal form $B' = A' + 1$ and $C' = B' + 1$.

   But, as in experiment II, we find that there exists an intermediate level (stage II) $N + k$ between the succession $N + 1$ and the numerical proportion. We also observe an intermediate stage III which is exemplified in the following:

PAL (7; 3) and VEL (7; 5) for 9 to C (question 3) give 3 to A and 5 to B because when you '*look at it, it seems right*' or because C '*eats a lot more*'; and for strips 2, 4 and 9, etc., they respond by measuring with their fingers.
   DIR (7; 4) for question 3, gives 5, 7 and 9: '*I take 2 away from 9 and 2 from 7*'.
   SYD (5; 10) for question 3 (9 to C): 3 to A and 7 to B. For the strips, when A has 1, he gives 2 to B and 10 to C, etc.: '*I give the biggest to C*'.

We are thus close to a sort of proportionality but to one which is still 'hyperordinal' or extensive. Metric proportions are reached in two steps: first (stage IV A) for one of the two relations without the other, and the second (IV B) for both. Here are examples of these progressive or immediate attainments of proportionality:

GOL (7; 6) successfully answers question 1 and where A has 2, he first proposes $B' = 3$ and $C' = 4$, then changes to: '(B) *receives 4 because it's two times 2 and (C) receives 5*'. For 9 to C, he uses the same reasoning but when asked to choose between 9, 8, 6 and 9, 6, 3, he understands: '(B) *is twice fish (A), it thus receives twice as much and (C) is three times the little one: it thus receives three times as much*'. When the strips are involved, he says the same thing. '(B) *receives twice as much as (A) and (C) three times as much*'.

HUB (8; 6) gives 2 to B and 3 to C if A has 1 (question 1). If A has 2, B will have 3 and C will have 4 '*because before there were 2 and 3. No, it won't work: 4 for (B) and 6 for (C) because (B) eats double and 2 times 2 is 4, and then for (C), I multiply by 3.* – And if (B) gets 6, how many does (C) get? – *9 because (B) has two times (A) and then (C) has 3 times (A), which makes 9*'. Strips: if B has 4, '*then (A) has 2, (B) has 2 plus 2 and (C) 2 plus 2 plus 2. – And 9 for (C)? – I divide by 3 to find (A) which makes 3 and I multiply (A) by 2 for (B) which makes 6 for that one*'.

MAR (8; 5) answers everything correctly saying: '*Because (B) is as big as two (A)'s and (C) is as big as three (A)'s*'.

The passage from this sort of preproportionality, simply involving the ranking of sizes, to the numerical proportion based on their relationships is thus gradual. The distribution (in absolute numbers) of the four stages thus described is as follows:

Age (years)	Beads ('pellets')					Strips ('biscuits')				
	I	II	III	IV A	IV B	I	II	III	IV A	IV B
5	5	1	3	0	0	6	2	1	6	0
6	3	1	1	1	0	3	3	0	0	0
7	1	2	3	1	0	2	1	2	2	0
8	0	2	0	2	3	2	0	0	2	3
9	0	0	0	0	5	0	0	0	0	5

We can see the slight progress indicated in the answers to the questions dealing with the discontinuous units (pellets) in comparison to the strips, whose numerical values are less immediately accessible. We observe above all a clear progress in these results in comparison to those of the second experiment (§2) in which metric proportionality is attained only at age 10–12. This is in part attributable to the fact that these 34 subjects were from an international school where the level of students is decidedly superior to that of other elementary schools in Geneva, but also no doubt due to the difference in the problems. In the second experiment, it was a question of prolonging a serial correspondence where the invariable beginning (the elements of the models) could remain insufficiently structured and thus be completed in an additive + 1 form, while in the present case the relationships $A = 1, B = 2A, C = 3A$ are presented in a variable fashion and more rapidly impose a transfer onto the corresponding A', B' and C'.

4. CONCLUSION

In the light of the two preceding experiments we cannot help making an interesting observation regarding the passage from the constitutive function of simple application to the constituted function of proportionality. Both begin from an assimilation of row II onto row I in the sense that the subject applies to the construction of row II the scheme of action which seems to him to be constitutive of row I, without any transformation of I into II. This results in the initial constitutive one-to-one function. But as Grize insists in his theoretical review, this assimilation already comprises a sort of 'pre-proportionality'. In general terms, relevant to classes as well as to ordered or seriable sets, this preproportionality comes to be the same as observing that for two sets A and A' possessing the same characteristics a or equivalent properties a and a' we have: a' is to A' as a is to A. In the case of seriations or ordered sequences, if $A < B$ and if the correspondence or application is expressed by $A' < B'$, the preproportion will be: B' is to B as A' is to A. We are here already dealing with quantities, since B comprises 'more' than A and B' 'more' than A', these quantities remaining somewhat quasi-qualitative or, more precisely, intensive, in the sense that their sole distinguishing feature is their order: 'more' after 'less' (regardless of the values involved).

But this order, which may be expressed in terms of ordinal numbers (first, second, etc.) quickly leads to a numerical correspondence based on the property which characterizes the finite whole numbers and which is evident as early as some of the preoperatory constructions of numbers, i.e. to every whole ordinal number there corresponds a whole cardinal number and vice versa. This leads to the intuition that, to A, B, C, etc., there correspond the cardinal values 1, 2, 3, etc. Since the differences between the ordered or seriated amounts are not as yet taken into consideration and the latter differ only in being 'more' or 'less', the proportion indicated at the time thus leads to the idea that these differences, already determined, remain equivalent and equal to + 1.

In the second experiment (§2) it is no doubt this immediate and false implication which leads the subject to neglect to make a detailed comparison of the intervals between the elements $A, B, C \ldots$ of row I and $A', B', C' \ldots$ of row II. Here the subject concludes that since there is 'a hole missing' in II, it will be enough to add $+1$ to the last term of row I to obtain the corresponding term in II. In the third experiment, the subject only expresses the ranks of the quantities in I in terms of $B = A + 1$ and $C = B + 1$ on row II which is to be constructed. But from ordinal preproportions the subject arrives quite naturally at other more complex ones which are not yet metric but which derive from scales which Suppes aptly calls 'hyperordinal'. On an ordinal scale, the subject simply posits that $B > A$, $C > B$, etc., without knowing by how much. On a metric scale on the other hand he knows the intervals or differences a between A and B, b between B and C, etc. (which remain indeterminate on the ordinal scale), and can express them in terms of units: $b = na$, etc. The hyperordinal scale is found mid-way between the two: although the intervals a, b, c, etc., can be compared, this can only be done in terms of 'more' or 'less'. The subject will know for example, that there exists a difference b between B and C which is larger than a between A and B, and if they are equal, he will only know that they are equal without knowing it in terms of units (i.e. without measuring them). The children in stages III and IV A of this experiment or in stage II of the preceding one (except for the fact that they express $N + K$ in terms of the number of holes) are at this level and their preproportions are already in the form[12] b' is to b as a' is to a, this form naturally lacking the equality of the cross-products.

Finally, the passage from there to constituted functions with a structure of actual or metric proportions is eased by the numerical quantification of the intervals and by the fact that in this particular case these numerical relationships are elementary and the solution of the problem is dictated by the presentation of the givens.

Lastly, it should be noted that during this passage from simple functional correspondences to proportionalities, the experiment in §3 presents us with a veritable morphism. At this level, functions go beyond simple correspondences to assure a transport from the structure of the original set. It is this transport of structure which in this case leads to proportions, and it is no doubt due to the direct nature of such a transport that proportionality is, in this particular case, attained earlier than would ordinarily be the case.

NOTES

[1] With the collaboration of F. Orsini (§1), M. Meylan-Backs (§2), and H. Sinclair (§3).

[2] In the sense, for example, of a 'group' whose internal compositions are closed within the group.

[3] For the details of the technique used and the general results, see Orsini, F., 'Régularités et systèmes des relations chez l'enfant', *Cahiers de psychologie* (Soc. de Psych. du Sud-Est), **VIII** (1965), 143–155. See also Orsini, F., 'A propos de l'étude de quelques régularités "naturelles" ', in *Psychologie et épistémologie génétiques, thèmes piagétiens*, Dunod, 1966, pp. 149–158.

[4] In the terminology of combinatory logic, C is defined by permutation. But there are three degrees: C_0 when a term B is simply substituted for A (as in the examples given here), C_1 when B is substituted for A and reciprocally (properly called a permutation) and C_2 when a set BA is substituted for AB, which we will call an 'inversion'.

[5] We also note, towards age 11–12, sequences such as *ab, aabb, aaabbb*, etc.; or *abb, aabbb, aaabbbb*, etc.; or even *aaaab, aaabb, aabbb*, etc. (In this last case there is a passage to compensations: see below.)

[6] The subject is given no other directions other than to complete the sequence which he does as he wants.

[7] It is no doubt for this reason that in the experiments on conservation at the preoperatory level, the subjects only take into account a single ordinally evaluated dimension (height of the level of the water, length of a sausage, etc.) which results in their systematic non-conservations.

[8] The following experiment was conducted by H. Sinclair as part of a series of experiments by S. Roller and M. Denis on proportions. We thank H. Sinclair for authorizing us to use it here.

[9] Piaget, J. and Szeminska, A., *The Child's Conception of Number*, Routledge & Kegan Paul, London, 1952.

[10] It should be noted that we first make them order the strips on a scale of 1 to 10 to facilitate a choice.

[11] For the definitions of intensive, extensive and metric quantities, see Chapter 14, end of §4.

[12] We find an indication in this direction in Stage I of the present experiment, but as we say, under the probable influence of causality.

CHAPTER 4

AN EXAMPLE OF CAUSAL AND SPATIAL FUNCTIONS[1]

If we assume that functions express the links inherent in schemes of actions in their dual causal and operatory or preoperatory aspect, it is reasonable to study in this respect a causal situation in which the variables are relatively easy to distinguish and to analyze both separately and together. A transitive movement resulting from a single thrust linked to an action, whether direct or instrumental, is certainly the most primitive example from the genetic point of view, but it is not very suitable for a detailed analysis of the functions involved given the global nature of their causes and effects as well as the difficulty experienced by young subjects in quantifying them with any precision and in particular in quantifying them metrically. We have therefore devised an apparatus which appears to be very complicated but is actually fairly simple. It is composed of weights z which stretch a spring[2] of variable length x through a string of constant length whose two segments y and y' are perpendicular and thus vary in an inverse relation (Figure 8). Additionally there are two arrows Fa and Fb, whose displacements, indicated on scales, correspond to the variations in the weights z and the length of the spring x.

The functions which enter into play in this apparatus naturally comprise, on the one hand, a causal aspect: such is the nature of the dependence between the lengthening of the spring x and the increase in z. But the

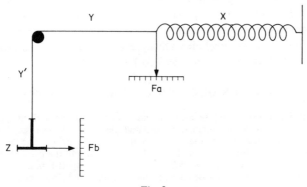

Fig. 8

emphasis can also be placed on the spatial covariations of the lengths x, y and y' without the displacements of the arrows Fa and Fb being taken into account, and the spatial variations can thus give rise to an operatory treatment, i.e. the conservations of $(x + y)$ and of $(y + y')$, the compensations between the increase in the lengths x or y' and the decrease in y, etc. The first problem will then be to examine the relative importance, by age, of the causal or spatial components, whether these be considered separately or interdependently.

A second question, dependent on the outcome of the above, will be to establish whether the functions involved are immediately composed into a total system, and if so, of what nature; or if the 'first' functional units are constituted by pairs, such as (x, z) or (x, y), etc. If this is the case, it will then be a question of analyzing the progressive composition of pairs among themselves, or more generally, the mode whereby dependences are linked among themselves. Does the pair (y, y') depend on the pair (z, x) or are these dependences linked together in another way? And above all, what is the general mode of composition of the dependences? If the variable v_1 depends on v_2 and if v_2 depends on v_3, does v_1 also depend on v_3? Are the connections comparable to transitivity (does this occur from the outset or later) or are the links more complex?

In fact, the links involved in the apparatus are circular and as such are fairly complicated: the force of weight z pulls on the segment y' which pulls on y, which in turn pulls on spring x, yet it is the stretching of the spring which makes possible the displacement of y from right to left and of y' from top to bottom, which in turn leads to the descent of weight z. While children of a certain level successfully distinguish these two directions of orientation of the dependences, it goes without saying that the younger subjects do not even consider them as long as they do not question the means or intermediaries by which weight z acts on spring x. However, these complications in no way exclude an analysis of progressive compositions since the latter are in fact mastered by age 9 to 10.

1. TECHNIQUE AND GENERAL RESULTS

To better distinguish the string segments y and y' of the above described apparatus, we placed two color strips behind them, one green and the other red, thus enabling us to call the segment y the 'green string' and the segment y' the 'red string'. The two arrows Fa and Fb are also of different colors. The scales which permit the measurement of the displacements of the arrows are calibrated in simple spatial units equivalent to centimeters by means of small cardboard fruits or vegetables.[3] This makes it possible to ask whether the arrow would go 'up to the carrot', etc.

We begin by presenting the material, making the subjects name the elements and allowing them to handle each one separately (the rest being hidden) so they can determine their role. Once everything has been made clear, we ask the subjects to anticipate what will happen to each of the elements when weights are placed and we ask in detail about their anticipations for each increase or decrease in weight. We then proceed to add the weights and request a description and explanation of the relationships observed. Two further questions are asked, one on the constancy of the length of the 'path' $x + y$ (the horizontal distance between the two screws, one of which secures the spring while the other allows the passage from y to y'). The other question involves the conservation of the length $y + y'$. Furthermore, in order to determine the degree of comprehension of the links, we suggest that the moveable arrow Fa is a mouse and that the second screw (which divides y from y') is a piece of cheese and then ask how we can help the mouse get his cheese.

Finally we proceed to comparisons between the displacements of the arrows Fa and Fb as we add or subtract weights (followed by a prediction of the weights needed to obtain such a displacement). This first of all involves the question of the existence, whether anticipated or accepted after the fact, of a link between the movement of the arrow and the weight,[4] then of the relation between the displacements of Fa and those of Fb. For this purpose, we can use either the qualitative scale of fruits and vegetables, a quantitative scale (in widths of 1, 2 or 3 fingers or in centimeters) or even make the subject construct his own scale.

The results of the first stage (age 4 to 6) can be summarized as the establishment of an initial pair between x and z, i.e. the stretching of the spring as a function of the weight. Here however either the intermediary variables are ignored or fairly unstable and uncoordinated pairs are constructed on the basis of observations, except (when the movements of the string are noted) if the movement is in the direction x, y, y'.

Between ages 7 and 10, we distinguish a second stage marked by a quantitative composition which is increasingly adequate for the functions involved, especially as regards the conservation of the lengths $(x + y)$ and $(y + y')$ and the compensation of the decreases of y through the increases in x or in y'. Similarly this stage is marked by the establishment of a relationship between the displacements of Fa and Fb, and a comprehension of their equivalence as well as of their functional link with the weights. There is also a third stage (age 11–12) which adds nothing new between the displacements and their weights, other than a numerical proportionality expressed in terms of double or triple, etc.

In order to provide a frame of reference, without attributing undue value to these data due to the number of intermediate cases and modifications made during the question period, here are the reactions of 72 subjects aged from 4–5 to 11–12, regarding the links between the spring x (whose stretching is acknowledged by all the children) and the string $y + y'$. When asked to anticipate, the subjects in group A, foresee the movement of the spring and the immobility of the string; the subjects in group B anticipate the movement of the latter but do not properly relate the segments y and y', even after having observed them; and the subjects of group C arrive at compensations. In the following table, the reactions to the question about the mouse and the cheese are found in parentheses: for group A no solution, for group B pulling on the string and for group C pulling on z or adding weight:

	Age 4–5	Age 6	Age 7	Age 8	Age 9	Age 10	Age 11–12
A	6 (8)	2 (2)	0 (0)	0	0	0	0
B	7 (4)	3 (3)	3 (1)	3 (4)	0	1	0
C	1 (2)	6 (6)	7 (9)	7 (6)	11 (11)	7 (8)	8 (8)
(No. of subjects)	14	11	10	10	11	8	8

As regards the conservation of the length $x + y$ and the relations between the displacements of the arrows, the successes are:

	Age 4–5	Age 6	Age 7	Age 8	Age 9	Age 10	Age 11–12
$x + y$	1	3	7	8	11	8	8
Arrows	1	2	8	8	11	8	8

The conservations of the length $y + y'$ are given under C of the preceding table.

2. CAUSALITY AND THE STAGES OF ITS DEVELOPMENT

It should be noted that the preceding technique did not provide for any questions of causality. Since causality is complex in the apparatus used while the observable elements are simple and their functional relations evident, it seemed methodologically practical for us to simply accept these laws without alluding to their explanation. However the facts have shown us that while the functions involved can be discovered by simple observations, they cannot be composed nor even predicted (in this experiment) without a causal interpretation, i.e. to understand the observations, one must be aware of the causal interpretation. We therefore tested twenty or so subjects aged 5–8, using the three following devices (Figure 9): (I) A red elastic (R) attached to

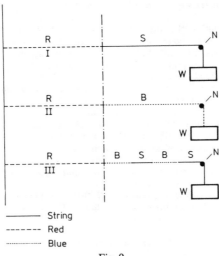

Fig. 9

a string (*S*) which hangs from a nail (*N*) and from which are hung weights (*W*) (rings *W*); (II) A red elastic (*R*) attached to a blue elastic from which are hung weights (*W*); (III) A red elastic (*R*) followed by four segments linked together: blue elastic, string, blue elastic and string with weight, the whole being the same initial length as in I and II. The questions involve the prediction of movements and their transmission (the role of intermediaries).

As we were able to deduce from the preceding results, the comprehension of the role of intermediaries and of the order of their succession presents fairly systematic difficulties until about age 7:

CRI (5; 1) in I predicts the stretching of *R*, then observes: '*The red elastic came up to there because we pulled*'. It is the weights which pull *R* and *S*. In II, same reactions. 'Are the rings pulling, (*R*) or (*B*) first? – *That one (R)*. – Why? – *Because it is shorter (R < B)*'. We return to *S* asking what the weights 'pull first? – *That (S)*. – Why? – *Because it is lighter*'. – In III, the weights pull '*that one (R)*. – Anything else? – *Yes, that (B + S + B + S)*. – Do those (the strings, etc.) pull the elastic (*R*)? – *No, because they don't go back up*'.

NAT (5; 1) explains the stretching of the elastic in I '*because it is taut*. – Is there something that makes it so taut? – *Yes, that* (the cardboard frame). – And what about that (the weights) do they do something? – *Yes*. – And that (*S*)? – *No*'. After that Nat states that any element in I, II and III 'pulls' any other one, without any order: in II the elastic *B* pulls *R* and vice-versa: 'And if we remove the weights, does (*B*) pull (*R*) again? – *Yes* (we then remove the rings). *Yes, it pulled it to there*': she shows the retreat to the demarcation line, 'pulling' here meaning to bring back to this line. At a given moment, the weights left in I seem to touch elastic *B* in II, and Nat then says: '*But it's not those* (weights in II) *which pull* (*B* in II), *it's that* (weight in I) *which pulls that* (*B* in II)!'.

AMB (5; 0) simplifies the problem by saying: '*You are the one who pulled* (*R*). – With what? – *With your hand.* – And that (the weights) do they pull (*R*) too? – *No.* – And does that pull (*S*)? – *No*'. – Then she admits that '*the rings pulled* (*R*). – And does the string pull it (in I)? – *No.* – And the blue elastics (in II), do they pull it? – *No.* – Then what makes the red elastic stretch like that? – *The rings*'.

ANA (6; 5) predicts in I that the rings will pull *R*: '*It will pull on that*. – And on the string too? – *No*'. Upon observation, she maintains that this is so. In II, the weight pulls *R* and *B*. We come back to I and Ana continues to believe that *W* does not act on *S*. – But how does (*W*) pull (*R*) without touching it? – *Because it's heavy*'.

TAL (6; 7) begins by hesitating in I: '*It pulls the string then the elastic. It pulls the elastic and then the string.* – What does it pull first? – *First the elastic and then the string.* – Why the elastic first? – *Because it's first*'.

BEN (6; 7): '*That* (*W*) *first pulls* (*R*) *and after that pulls* (*S*). – First (*R*)? – *Yes.* – Why? – *Because it's last!*'

CAR (7; 4) in I: *W* pulls *R* and also pulls *S*: 'First on (*S*) and then on (*R*) or first on (*R*) and then on (*S*)? – *First on the elastic.* – Why? – *Because it's heavy. That one* (*R*) *is the first one and that one* (*S*) *is the second*'.

There is therefore no doubt that the order of the transmissions between the weights and the red elastic poses a constant problem and that even when the order seems to be respected (Cri in I when we go back to it) it is for reasons which are unrelated to the succession of bottom to top: because *S* 'is light', with the succession itself being denied because 'it won't go back up'. And yet in the case of instrumental pushing (hand → stick → external object) the transmission of the push is accepted and well ordered by subjects of this age group (5–6 year-olds). Pulling on the other hand, remains obscure because in this particular case the objects are displaced from left to right and from top to bottom, while the pulling force is transmitted by intermediaries from bottom to top. In the transitional stage between I and II, the subjects appear to understand, but encounter the same difficulties and conclude that it is a simultaneous general effect:

NIK (5; 2) does not predict anything in I, but after observation, correctly concludes: '*the rings are heavy, the string is attached to the elastic, thus the string pulls because of the rings*'. But he finds it impossible to determine an order: '*It pulls both at the same time*'. In situation II, same reactions: 'Do the rings pull first (*B*) then (*R*), or first (*R*) then (*B*)? – *Don't know. Both at the same time*'. The order in III is not understood any better.

The subjects of stage II establish the order of succession of the intermediaries without difficulty, for the most part from age 7–8 and sometimes even earlier:

MAR (5; 8) does not predict anything for situation I but then understands right away: '*Because it is very heavy, it makes the* string and that (*R*) move, because the elastic is attached to the string. – What do the rings pull? – *The string.* – And who pulls (*R*)? – *It's the string'.* Situation II: *ibid.* In III, she predicts that the elastic will not be pulled *'because there are several* (segments)', then realizes: *'It's because it is very heavy.* – What do the rings pull? – (shows the segments in ascending order)'.

BO (6; 6) predicts well for I: *'It's heavy, it makes the elastic stretch.* – But the rings don't touch it. – *It pulls the string which in turn stretches the elastic'.* Situation III: correct order.

PHI (7; 4). Same reactions. Situation III: *'The weights are attached to the string and the string to the elastic* (shows the correct order) *and so that pulls on it* (*R*)'.

We thought it would be useful to provide this data on causal comprehension because it not only clarifies but possibly even explains the reactions to questions involving functions asked during stages I and II to be discussed in §§3 and 4. As for stage III, it would be useless for us to comment here on the causal interpretations to be drawn since they will be made sufficiently clear (as we will see in §5) by the questions asked about the functions themselves.

3. STAGE I

Let us return to functions by citing the case of a 4 year-old subject who made no predictions and thus made it possible for us to observe how he established the initial links, after which we will present some less primitive cases which will show how successive pairs are constituted:

CEC (4; 5) predicts no movement when we place the weights but when faced with facts to the contrary she says: *'That* (*y*) *did not move, that* (*x*) *did.* – And if I remove the weights? – *It works.* – How? – (Gesture from right to left). *Bigger* (she removes the weights and sees the opposite)'. She also does not forsee that we can pull *z* instead of adding weights. We then replace these asking her to look at the string: *'Bigger* (*y'*). – And the string here (*y*)? – *Smaller'.* Her only solution for the question involving the mouse is to cut the cheese and she does not anticipate any movement of the arrows. As regards the 'path' *x* + *y*, *'it will change.* – Let us see (we place the weights). – *That makes it bigger* (it is thus a question of the distance shown between the 2 screws). – Why? – (She shows the weights)'.

FRE (5; 1) does not make any predictions either (we use an elastic instead of a spring) but sees that the weight stretches *x*: 'And the string on the red (we point to *y*)? – *Bigger, I think.* – Look. – *No, smaller.* – If I remove the weight, what happens to the elastic? – ... (he observes): *smaller.* – And if the elastic becomes smaller by that much, does the string (*y*) become larger? – *A little bit smaller'.* He doesn't find any answer for the question involving the mouse, but when asked what the effect of a weight would be, he finds that: *'The elastic will be bigger* (i.e. reach further) *and the mouse will be able to eat the cheese'.* Thus his observations lead him to discover the pair (*z, x*) in both directions, stretching and contraction, which he likewise immediately applies to (*z, y*), no doubt basing himself on the idea that a displacement to the left equals stretching.

VER (5; 7) knows that an elastic stretches but that the string does not. He thus predicts the links (*z, x*): *'it will happen later'* for (*x*), while for *y*: *'It goes, it falls, it *comes down'*. There is also a link between z and $y + y'$ (as a whole) since they both move down, but this is then equally expressible as a stretching: *'It* (the string $y + y'$) *became longer. –* (We try to make him dissociate y and y'). *–* But the red string there (y)? *– It became a little shorter. –* Why? *– Because it is heavy, so that* ($y + y'$) *advances. –* And the piece there (y')? *– Bigger. –* Does the string stretch? *– No. –* Well? *– Because it is heavy, it advances*. But the 'path' $x + y$ is not conserved and becomes larger, and the total $y + y'$ is not conserved either: y' stretches more than y shrinks. Aren't these the same paths (total $y + y'$ depending on the weights)? *– 'No, because there* (y') *it's heavier'*. By contrast, given this role of the weights, Ver succeeds in the problem of the mouse: *'You've got to place all of these things* (weights)'. He accepts the link between the weights and the arrows but does not anticipate that the displacements of Fa will equal those of Fb. This case therefore shows clear progress over the preceding ones, through the constitution of the pairs (z, x), $(z, (y + y'))$ and even (z, F) and through their juxtaposition according to the apparent order of displacements of x, $(y + y')$ and z.

PAT (5; 7) is at the same level as Ver, but since we are dealing with a spring, he anticipates to-and-fro movements and shows the possible stretching and contraction with his hands. The initial function is again $x = f(z)$, which is then applied to the string *'because it gets bigger'* ($y + y'$ as a whole). But if it stretches, it is *'because it is heavy'*: linking of z and ($y + y'$); and also *'because the spring has stretched'*, which yields a new pairing: $(x, (y + y'))$. He then observes that y *'will be much smaller . . . because it turns like this* (passage to y')', but there is no compensation between $-y$ and $+y'$ *'because that one* (shows the path $x + y$) *goes like that* (horizontal) *and that one* ($y' + z$) *like that* (top to bottom)'. The string y' advances more than a finger and y 'less'. The path $x + y$ no longer has a constant length. On the other hand, Pat succeeds in the question involving the mouse: *'You must place a weight since the string* (y) *becomes smaller'* which leads to a new pairing: $(z, -y)$.

ARI (5; 8). Same reactions: the pairing (z, x) leads to $(z, x + y)$ with non-conservation of the distance between the two screws, since the road *'becomes bigger. –* If we walk, etc.? *– There will be more to walk'*. *–* The entire string $(x + y')$ *'advances, gets bigger'*, in accordance with the same assimilation between 'further' and 'longer' evidenced in the responses of Fre, Ver and Pat, and by virtue of the ordinal estimates proper to this level, but reinforced by the idea that weight z stretches the string ($y + y'$), derived from the model (z, x). Contrary to Pat, who did not foresee any displacement of the arrow, Ari anticipates it right away (pairing of (z, F)) but does not equate the displacements of Fa and Fb. By contrast, the last three subjects admit the inverse functions immediately: if we decrease the weight, everything is shortened and the direction of the movement of the arrows changes. However, their failure to preserve the distances keeps us from speaking of inverse operations.

ALA (5; 11). Same reactions, but with regard to ($y + y'$) he states: *'The string becomes bigger, but not like the elastic'*, which is the same as saying that it becomes longer as a whole, but does not stretch.

GRA (6; 11) in spite of his age still believes that the string will *'become bigger'* because it descends with the weight and for this same reason the length of the path $x + y$ varies. He does not accept compensations between y and $+y'$ nor does he expect the arrow to move if the weights are modified. He is nonetheless convinced by the facts and by the observation of how the parts are attached. He then correctly predicts that Fb will come down with more weight and go up with less, but still assumes that Fa will go to the left in the case of a decrease in weight. Once he has clearly understood, we use only Fa and Fb as measures of the decrease in y and the stretching of y': 'then where does Fb go if I place this weight? *– There* (correct). *–* Isn't that the same path (displacement of the arrow) as here (Fb hidden) or not? *– Here* (Fa) *it's smaller and there* (Fb) *it's larger. –* Why? *Because there* (y', thus Fb) *it will go down and there* (Fa, thus y) *it will advance'*.

These reactions first of all show the importance of an initial function which translates the action observed by directly linking its point of departure (introduction of a weight) to its end result (movement of the spring or elastic), i.e. the pair (z, x). Its global nature goes without saying since the two intrinsic aspects of an action are its causal beginning and its end or terminal effect. The subject Cec thus says that the elastic 'moved' and does not even note that this is also the case with the string, which other subjects also see but completely ignore. This function even precedes the discovery of its direction and factual observations are needed to arrive at the form which will quickly become generalized: the stretching of x coupled with the increase in weight. The inverse appears shortly thereafter, but not immediately as we saw in the case of Fre.

The second fundamental initial function is of a much more general nature than the present situation would suggest, i.e. it comprises the correspondence between the order of the terms and their quantitative value. We saw good numerical examples in Chapter 3, §§2 and 3, where any increase is reduced to $n + 1$ due to the fact that the rank $n + 1$ is derived from rank n by the addition or subtraction of an ordinal unit. The equivalent in the case of spatial relations is the assimilation of 'further' to 'bigger' or 'longer'. This assimilation is thus not unique to the context studied here but it is considerably reinforced by the fact that, in this particular case, any displacement is the result of pulling.

Once the stage I subject has discovered the pair or function (z, x), he immediately applies it to the other variables and even to the constants. He does not believe that the string is elastically stretchable (Ala is very clear on this point), but merely that the weight pulls the string as it pulls the spring. In this case, it 'advances', it 'goes further' (Ver), etc., and by virtue of the aforementioned quantitative ordinal correspondence, the entire string $(y + y')$ becomes longer, the two segments not being distinguished: this leads to the pairing $(z, (y + y'))$ similar to (z, x) except that it lacks elasticity. The same applies to the path $x + y$: since the moveable elements (x and y) which travel along it are pulled by the weight, the path itself grows so that 'there will be more to walk' (Ari).

The idea that the string stretches could be inhibited by a notion of conservation but Vinh-Bang has shown (*Études d'épistémologie génétique* **XX**, p. 7) that a string of constant length which is divided into two perpendicular segments A and B is not conceived as conserving its length when the lengths of A and B vary, i.e. the stretching of A does not entail an equal contraction of B and vice versa prior to age 7.[5] In the present experiment where the string is pulled by the weight z, the same applies *a fortiori*: the subject is able to see clearly and then to predict that y will decrease and that y' will increase, but the pair $(-y, +y')$ involves no compensations and thus remains subordinated to $(z, (y + y'))$, then to (z, y) and (z, y') but without adequate composition of the lengths.

Finally, as regards the arrows, the process is the same but with a slight *décalage*. First, they are considered immobile (cf. the level of Cec for the string). Next, they are conceived as being moved by the weights and thus leading to (z, Fa) and (z, Fb), but these pairs are not related and thus the displacements are not equalized. The inequalities are again attributed to the pulling: Fb goes farther 'because it will come down' (Gre).

There then remains the problem of the composition of these pairs, all analogous by virtue of the application of the same functional scheme. The only composition which sometimes occurs in stage I suggests the direction of the movement and not the direction of the transmission of force exerted by the weights on the string and from there on the spring. It is not that this composition is erroneous, for the movement of the string depends on that of the spring and the descent and stretching of the segment y' depend on the displacement and shortening of y (these functions and their inverses thus being true). This is not however what the subject is saying. The results of Chapter 1 on the composition of pairs should make us proceed prudently, since it is only between ages 6 and 8 that 75% of the subjects are able to connect the paths when changing trains to go from point A to point B via one or two itineraries in a very simple and clearly visible network. Thus if this experiment had dealt with linking the spring to the weight by one of the several possible paths and not only by y and y', we would have found the same discontinuities as those displayed in Figure 3 (I to V B) for children aged 5–6. Consequently the only composition of pairs which appears more or less certain in the preceding responses and which is implied in the expressions 'to advance, to go down', etc. is that of the succession (z, x) before (z, y) and (z, y) before (z, y'), etc. These are not functions of functions, but simply a variable order of succession. (For this order see the responses of stage I on causality in §2.)

4. STAGE II

The two criteria of stage II are the conservation of the distance $x + y$ comprised between the two screws of the framework of the apparatus, and the conservation of the length $y + y'$ with compensation between the decreases in y and the increases in x or y'. Added to these is the almost

immediate comprehension of the equality of the displacements of the arrows
Fa and Fb, which is the same as saying the equality of the increase Δx and
$\Delta y'$:

PIE (7; 4) is an intermediate case between stage I and II. He understands that y' becomes
longer when y decreases in length, thus when y is shortened with the rest of the apparatus
hidden, he immediately grasps that '*you added a heavier weight*'. But when he is shown
(by means of a small stick) the length lost by y, he is not at first able to deduce that y'
will become longer by an equal length. When asked if he can find the 'end of the string
coming from y', he exclaims '*Oh! Here*' (in y') and then admits that the length $y + y'$ is
constant. On the other hand, due to the lack of transitivity, he still refuses to accept that
the increase in the length of the elastic Δx will be equal to that of y, or Δy: '*It will be
bigger, because the elastic stretches while that* (y') *just goes down*'.

ANA (7; 4) on the other hand together with the following cases is a typical stage II
subject. The distance $x + y$ seems constant to him because when the spring stretches
'*there is less string* (y) *and more spring*'. As regards what y gains, it is '*as much*' and not
less nor more than what y loses. Lastly, the displacements of arrows Fa and Fb are '*the
same length because they move at the same time: it will also push that one* (Fb) *when it
goes down*', thus $\Delta x = \Delta y$ '*as much here as there*'.

FLO (7; 4). The distance $x + y$ remains constant '*because if we add a weight it's the
same length*'. The same with $y + y'$ '*because the string is always the same length*'. As
regards the displacement of the arrows, '*if there are more there, it makes more here*'
because it's the same weight '*which pulls them*'.

JEA (7; 6). The path $x + y$ is constant '*because the spring replaces the string*'. The
arrows '*cover the same path: if you hang the weight, there is the string which is not
elastic*'. For 2 boxes in Fa we will have 2 boxes in Fb '*because the string always remains
the same length and the spring stretches the length of two boxes*'.

PAC (7; 6): $x + y$ '*is always the same path: the paths don't move, they can be
lengthened*'. As regards the arrows: '*If we place the same weight, they cover the same
path*'. He then predicts that Δx, Δy and $\Delta y'$, will be equal.

CUR (8; 4). Same reactions: '(Fa) *was stretched one finger. We could measure with
the weight which was placed. You placed 100 grams* (spontaneous!). – And if I place
200? – *That will make two fingers*'. And the same in Fb, in y', in y as in Fa and in x.

TRI (8; 7). Same reaction (except for the measurement in grams) but in the opposite
direction: if we pick up the weight placed '(x) *will also become shorter by 2 fingers, and
(y) will become longer by two fingers. It's very easy. –* But how are you sure of
it? – *Because that* (y') *goes up*'.

MAR (8; 7): the distances travelled by the arrows are equal '*because they move at the
same time*' and FRA (8; 11): '*Because the weight makes both move at the same time*'.

LOV (9; 5) states that an increase Δx in the length equals the displacement of Fb:
'How do you know? – *Because it's the string which pulls it and makes one box less*' and
FAL (9; 5): '*We place a weight, it pulls the string*', which leads to the equalities.

We thus see the generalized nature of these conservations and equalizations
which naturally constitute an aspect of operatory reasoning. But there is also
the causal point of view, and Jea makes us perceive its significance, however
implicit, when he declares:[6] 'If you hang the weight, the string is there, and it
is not elastic'. This is essentially the same as saying: the action of the weight
on the spring presupposes the instrumentality of the string, and since the

latter is of constant length, the lengthening (x) or the displacements $(y, y', Fa$ and $Fb)$ will be equal in length in the direction of descent as well as in the opposite direction. In other words, the transmission of the pulling action from bottom to top in turn entails the stretching of the spring and a series of equal displacements from top to bottom. It is the implicit comprehension of this fact that enables Cur, at age 8, to set up a spontaneous functional dependence between the values of the spatial displacements and that of the weights themselves. This results in a new mode of functional composition which is both causal and operatory. Thus, instead of a series of juxtaposed pairs such as (z, x), $(z, (y + y'))$, (z, F), etc. translating a set of analogous actions, each developed by itself, all the variations at this stage become covariations in relation to each of the others in both directions, i.e. the ascent of the actions and the descent of the successive effects, even though everything occurs almost instantaneously. The principle behind this composition of pairs differs from that of stage I in that in the latter there was a privileged pair (z, x) to which others were added one by one, whereas in stage II there is an immediate awareness of the intermediaries in both directions between the weights (z) and the spring (x): 'It is the string which pulls' as Lov and Fal state, which is the same as inserting the pairs (z, y), (y', y), (y, x) (as well as (y', Fb) and (y, Fa)) between z and x in the direction of the successive pulling actions (each element pulling the next one) and the same pairs in the reverse order $((x, y), (y, y'), (y', z)$, etc.) in the direction of the displacements.

5. STAGE III

Three new factors characterize this last stage: (1) a slightly more explicit formulation of the interdependences, (2) the statement that the spring is attached or restrained, thereby introducing an upper limit to this system of the ascent of the pulling actions and the descent of the displacements, and (3) an operatory progress in the structuring of covariations which become proportionalities:

HAM (10; 7) gives the correct equalities when he says '*because if we place a weight, it makes them advance at the same time there and there*', etc. He does however specify that while the weights pull everything '*the spring holds back the string*' and does not limit himself to explaining its descent.

GER (10; 10) likewise mentally travels the path in both directions: '*That goes down there* (z) *and that pulls it up*' and '*that pulls on the spring and that goes down* (y)'. He establishes a correspondence between the displacements of the arrows and the units of weight while stating that he does not know what they are in this particular case.

CHA (11; 2) states that the string *'does not stretch, it goes down'* with compensations between *y'* and *y* remaining *'the same length'* while the spring *'stretches because it is fixed'*. He then proceeds to construct a scale of spatial displacements which are in proportion to the weights.

MEI (12; 1) explicitly indicates both directions of the action: *'If I place a weight, the string is pulled by the weight and the force of the weight pulls the spring by that much* (the distance *Fa*) *after which the string* (*y*) *goes down there* (*y'*)'. But if we remove the weight *'it goes back up to where it was before, because the string is pulled again by the stretching spring'*. Then he too constructs a scale with exact proportions of *'double'* or *'triple'* lengths for the corresponding weights.

SAG (13; 1): *'When the string is pulled the weight will stretch the spring, which will make the string here* (*y*) *smaller'* and will make *y'* bigger. And *'when we add weights we realize by how much it went down'*, the displacements are always equal *'because if it goes down here* (−*y* and +*y'*) *the spring is stretched by the same length'*. Thus for a double weight we will have *'twice this* (unit) *length'* in *x* and *y'*, etc.

This stage is thus marked by the consideration of a new pair, introduced between *x* and its point of attachment, and explains why the spring 'stretches since it is fixed', while the string 'does not stretch, but moves down' (Cha). These facts were already evident in stage II but the reason why they are only now being made explicit is that the system which has become circular has been closed: the weight exerts force on the string (Mei) which in turn transmits it to the spring (Ger, Mei, etc.) which then stretches resulting in an inverse displacement of the string which pulls the weight, but 'restrains' the spring (Ger) and the descent of the weight (Ger). Inversely if we decrease the weight, everything goes up (Mei). This causal circularity is expressed by the proportionality of the weights and of the displacements, as each of these subjects confirms on constructing the proper scale of the distances covered and the respective weights.

6. CONCLUSIONS

This experiment first of all exhibits a function which, as an expression of the links proper to actions in general (in this particular case of actions performed by objects on objects), constitutes the common source of operations (e.g., equalities, conservations and proportionalities) and of causality (e.g., the total circular system made explicit in stage III). In effect, these functions essentially express a dependence and it is the composition of these links of dependence which will be oriented in either a causal or explicatory direction (the system as a system of material links) or in an operatory or implicative direction (the composition of equalities, compensations, etc. treated as geometric entities capable of being constructed or reconstructed through the actions of the subject).

Constituting the points of departure of this composition are those pairs which simply express the dependence between the two terms linked by the initial function: weight z pulls spring x, or $x = f(z)$, it being possible to rediscover this (z, x) pair in other elements such as $(z, (y + y'))$, etc., through the application of the same scheme. The first type of composition will thus be the passage from the simple juxtaposition of pairs to an order of succession (stage I), without the integration of the *interdependences* onto the elementary *dependences*. How does the subject proceed from this general lack of coordination to a composition assuring the interdependences?

The preceding facts are interesting in that they provide us with an example of a progressive composition where the interdependences are simultaneously constructed both causally and operatorily. From the causal point of view the two essential discoveries are: (1) the intermediaries between cause and effect, and (2) the possible inversion of the entire set of actions. From the operatory standpoint there are two constructions corresponding to these: (3) the transitivity of the dependences and (4) compensations or conservations.

(1) The causal intermediaries appear in stage II (because even when stage I subjects sometimes perceive them, they do not take them into account)[7] and are made explicit in stage III: the force of weight z is exerted on the spring, by means of the string $(y + y')$ and in return the spring fixed at x' is stretched Δx, resulting in the equal displacement Δy (and Fa), $\Delta y'$ and the descent of the weight as measured by Fb. The necessary intermediaries, from the causal point of view, thus end up by replacing the juxtaposed or simple successive pairs (z, x), $(z, (y + y'))$, (z, F), etc. by the pairs (z, y'), (y', y), (y, x), (x, x') in the direction of the ascent of the pulling actions and (x', x), (x, y), (y, y') and (y', z) in the direction of the descent of the displacements (including (y, Fa) and (z, Fb) in both directions). These pairs are then conceived as interdependent and are not limited to being just a series of dependences.

(2) This structure of totality is such that it remains well-structured even when its order is inverted. Beginning with stage I, although not immediately (cf. Cec and Fre), the subject comprehends certain local inversions (as with Fre but only after observation) such as the contraction of the spring when a weight is removed. In stage II, on the other hand, this inversion of the order of actions, (which is not identical with but corresponds to operatory reversibility, the reversible being distinguished from the invertible) becomes general in both directions of the process, i.e. if we take away the weights, the pulling force decreases in the ascent, the spring contracts and, in the downward direction, y becomes longer, y' decreases in length and Fb goes up.

The pairs of dependences remain the same, but the differences Δx, Δy, etc. are oriented in opposite directions.

(3) To the necessity of the causal intermediaries, thus to the material interdependences which are the source of the concatenation of the pairs, there corresponds, from the logical or operatory point of view, a certain transitive composition of these functions. Of course, the relation of causality is not transitive in a complete sense: if a is a cause of b and b is a cause of c, then a is simply a condition of c. On the other hand, in terms of functional dependences, if b depends on a and c depends on b then c also depends on a which unites two simple functions into one composite function. From such a point of view, the coordination of the functional pairs can take on the structure of a kind of operatory seriation.

(4) Above all, the material invertability (*tollitur causa . . .*) will take the form of operatory reversibility, thus of a system of compensations: if y decreases and y' increases when a weight is added, and if y' decreases and y increases when a weight is removed, both of these covariations being concomitant in both directions, the subject concludes that $\Delta y' = \Delta y$ because the general interdependence of the two covariations suggests this equalization. Since it is only a question of displacements, the probability of this equality becomes increasingly evident until the subject is finally able to understand that what is lost by one of the segments y or y' is necessarily gained by the other. Thus these compensations result in conservations.

Finally, this progressive composition in its two complementary aspects, causal and operatory, leads necessarily to a system of proportions comprising three stages. In stage I, there is already post-observational comprehension of the fact that if x increases, y decreases or y' increases. Although the amounts are not yet known, this is already a preproportionality (a correlate as used by Spearman), in the sense that the initial state of x is to that of y or y' as the final state of the former is to the final state of the latter. In the second stage, under the influences of the compensations and conservations, the differences are equalized: $\Delta x = \Delta y = \Delta y'$, etc. But since the incorporation of the interdependences into the elementary dependences requires the establishment of a functional relation between the weights and distances (more precisely between Δz and Δy, $\Delta y'$ or Δx) which results in the metric generalization of the equalities, i.e. in the numerical differentiation of the weights on the basis of their distinguishing effect on the distances (in x, y, y' or F). Thus sooner or later (as soon as the operatory instruments are available) proportionality results from the fact that the interdependences end up by encompassing all the dependences.

NOTES

[1] With the collaboration of M. Meylan-Backs and A. Papert-Christophides.

[2] Or an elastic for some young subjects.

[3] Fruit for one of the arrows and vegetables for the other, so as not to suggest the equality of the displacements.

[4] It should be noted that the arrows are fixed to the support of the weights and to the point of attachment of the spring and the string in a manner which imposes the perception of a material link.

[5] See also below, Chapter 8, note 1.

[6] In §2 we saw the reactions of stage II to this.

[7] In §2 we saw how this problem of intermediaries remains basically unresolved in stage I.

CHAPTER 5

FROM COPROPERTIES TO COVARIATIONS:
THE EQUALIZATION AND ESTIMATION OF
INEQUALITIES[1]

The function which we will here study can intervene in a causal context (as in Chapter 4) as well as operatorily. Let us suppose that in the preceding experiment, the segments of strings Y and Y' were equal at the outset, each one being 5 units long such that when Y' is made one unit longer the difference between Y' and Y would be 2 (= 6 and 4), and if the same action were repeated, it would be 4 (= 7 and 3), etc. In this chapter, we have a set of marbles rolling in a slide and falling off one at a time from point C. If at a given moment there are 5 marbles still on the slide and 5 which have already dropped to the lower level, the subsequent drops will yield 4 and 6, 3 and 7, etc., respectively for the two collections. This variable difference of 0, 2, 4, 6, etc. resulting from each successive displacement of one element from the first collection to the second can be problematic for young children and it is this aspect which we intend to study. Let x be the number of transfers of n elements ($n = 1, 2, 3$, etc.) and y be the difference between the two collections. Then the difference to be determined is $y = f(x)$ and in this case $y = 2x$.

But we will pose the problem in an operatory context, i.e. one in which the collections of objects, their transfer from one collection to the other, and the resultant sums and differences derive only from the actions of the subject. From the causal and, in particular, spatial aspects of these actions, we can dissociate arithmetic links which can even be established without any external movement. We thereby hope to encounter this function in its simplest form (which does not mean the most elementary) from the genetic standpoint and to thus be able to analyze its progressive constitution.

1. TECHNIQUE AND GENERAL RESULTS

We place 1 to 4 tokens before the child depending on his age (all of the tokens are of the same size and color, including those of the experimenter), and then place the same number before the experimenter. Once the child has agreed that both collections are equal, we say: 'Now I am going to play some tricks on you. I will take some of your tokens and hide them under my hand together with mine. But since you still want to have the same number as me, you will have this little box (an open box full of tokens) from which you can take as many as you want so as to always have the same as me'. We begin by taking away one token and hiding it, in plain sight of the subject, under the screen (the experimenter's left hand which is already covering the other elements). The subject thus sees that we have added one token to those which were already there (without counting them again) and that he must simply take from the reserve only the number needed to equalize the two piles. We continue likewise with 2 or 3 tokens up to 5, depending on the subject's level.

Next, depending on the case, we ask three control questions. The first is asked during the experiment: instead of starting with two piles of 4 from which we will take away 1, 2 or 3 tokens, we skip for example to two piles of 10, to see if the method used is different. After 1 to n tokens have been taken from the subject's pile, the second control test consists in continuing or starting over again, this time putting the tokens taken away from the subject into the reserve box instead of into the experimenter's pile. The function $y = f(x)$ thus becomes $y = x$ with the subject's adaptation to this new situation indicating whether or not he understood the preceding one. Thirdly, if the subject has reacted correctly to the main question several times in a row, we then ask him the following without any new manipulations: 'If you take n tokens from the box (for example if you take 2), how many must I take from you so we each have as many as the other?' (here 1). Thus this question involves an inverse function, such that, starting with $y = n$, we find $x = n/2$. If the subject has difficulty understanding this question, we simply ask him: 'If I take a token from you, why must you replace two?' The experiment generally ends when the subject reaches the limit of his current capability.

We can distinguish four stages in the results of this experiment. Let y be the number of elements needed to equalize the piles and x the number of elements transferred from pile A (child) to pile B (experimenter). The first stage (age 3–6) comprises two substages I A and I B: in I A we have $y = x$ except for very small piles (1 or 2) and in I B we have $y = nx$, n being variable without comprehension nor even regularity. After varied intermediate reactions we encounter stage II which is characterized by the subject's attainment of certain regularities, although he does not understand them. It is the degree of this non-comprehension among subjects aged 5–6 that the control questions serve to determine. In stage III, we find the progressive and inductive discovery of the law and in stage IV (from age 9–11) its comprehension and explicit deduction.

2. STAGE I

The earliest level at which this experiment can be conducted is naturally the one at which the child is capable of correctly judging by sight whether or not two piles are equal. To find this level, we worked backwards to age 3, where the limit was 2–3 elements. Here are some examples of the first level (I A):

PIE (3 years) is able to reproduce a row of up to 4 elements. We give him 2 tokens ($A = 2$), and keep 2 ($B = 2$) then take away 1 token ($x = 1$). He takes 1 from the reserve

(y = 1) without wondering why B becomes 3. On the other hand, for $A = B = 1$ and x = 1, he takes y = 2 because he understands that if he has no more tokens in front of him (A = 0), we must have both. Yet if we start over with $A = B = 2$ he again takes just y = 1 which results in A = 2 and B = 3. When we take away 1 more he takes y = 1, thus leaving A = 2 and B = 4. 'Do you have the same as me? – Yes'. We start all over again with $A = B = 1$ and he repeats the same actions.

ARL ($3\frac{1}{2}$ years), for $A = B = 2$ takes y = 1 several times for x = 1 until A = 2 and B = 4, then A = 2 and B = 6 without realizing that B increases. We begin again, but for $A = B = 1$, in contrast to Pie, he only takes y = 1.

LUC (4 years) for $A = B = 1$ is, on the other hand, able to understand immediately that y = 2. But for $A = B = 2$, he takes y = 1 and continues up to B = 6.

DUR (4; 9) when starting with $A = B = 2$ takes y = 1 each time until B = 5. But when we uncover the pile, he is very surprised and adds 3 to A. 'And now (x = 1, thus B = 6 and A = 4)? – There (y = 1). It's the same thing. – (We uncover it again). – No, there|(he adds 1)'. We continue likewise until B = 12, each time displacing x = 1 and each time he adds y = 1, showing the same surprise every time he sees the inequality.

We start over with $A = B = 4$ and he reacts similarly until B = 12 even though we uncover pile B every two times. When we start over with $A = B = 3$, he again follows the y = 1 method, but when B = 7, after several observations, he places y = 2 for x = 1 twice in a row. 'Why? – When I need it, I take 2. – When do you need it? – When you say so (= when you show the result). – Look then (we displace x = 1). – I put one in its place (y = 1). – (We uncover). Oh, no!' but he continues in the same manner until B = 14. 'How many do I have? – We each have 8'.

FRA (5; 3) reacts like Dur indefinitely. When we uncover pile B, he is very surprised but doesn't understand why: 'Oh! – What did you need to do to have the same thing? – I don't know'.

Aside from the successes of Pie and Dur when $A = B = 1$ (which is not even generalized as shown by the case of Arl), we can see that these subjects do not understand that a transfer of elements from A to B increases the collection B at the same time that it decreases the total A, i.e. they consistently take only the same number of tokens as were taken away from them (here $y = x = 1$ constantly) and are convinced that they have maintained the equality $A = B$. When asked to verify this, they immediately correct themselves but do so without understanding the reason nor conserving the method of correction. In short, the function involved here is only the application by bijection of A onto B and x onto y, without any thought to the variation of B.

In substage I B, which bridges the gap between stage I and stage II (itself characterized by intermediate but somewhat more stable reactions), the subject sometimes gets the feeling that pile B increases, in which case he takes $y > x$. But since he does not understand the relationship between x and y, these inequalities are sporadic and transitory:

TIS (4; 6) for $A = B = 2$ with x = 1 starts by taking 1 token: 'Is it the same, like that? – Yes, no (he replaces 1). – Is it the same? – Yes. – Sure? – No (he removes 1). – Is it the same? – Yes, because I have 2 tokens (A). – And me too? – Yes ... no (he adds 1). – Sure? – *Yes.* – And now (we take away $x = 1$). – *There* (he takes $y = 1$), *that's the same as you have.* – And now (we take away 1). – *I take 2.* – Why? – *Because you have many'.* We continue but he still only adds $y = 1$. We then uncover the pile ($A = 10$ and $B = 4$): '*Oh!*' We start over with $A = B = 5$ and take $x = 1$: '*I take 1.* – Why? – *Because you have 5* (he forgets x). – And now ($x = 1$ and $y = 1$) how many do I have? – *Five.* – And now (new $x = 1$)? – *I place 2 because you have many'.* But then he continues to put $y = 1$ until there is a new verification: '*Oh!* – Look here (we begin again at $A = B = 2$ and $x = 1$). – *I replace 1, only one. That way you have as many as me.* – How do you know? – *I know how to count'.* We start over with $A = B = 1$ and he replaces $y = 1$ after his has been removed. He continues several times with $y = 1$, then suddenly says: '*No, that again* ($y = 3$) *because you have that many'*, then some $y = 1$ and again suddenly $y = 4$ for $x = 1$: '*I know how to count, I know that you have a lot* (there is in fact a difference)'.

FRI (4; 11) begins with $y = 2$ for $A = B = 1$, then passes on to $y = 1$ for some $x = 1$. When we uncover B he is very surprised and during the following displacements (all $x = 1$), he takes $y = 2$, then $y = 3$ and is ready for $y = 4$ but asks to look when he places $y = 3$ (at that point collection A has 10 elements and B only 7!).

JOL (4; 11) reacts like Fri but after having refused to make any prediction ('*I don't know any more, let me see'*), he adds $y = 2$ then $2 + 5$ and ends up with $A = 10$ and $B = 5$!

These subjects begin by reacting like those at level I A, but what is new and important is that they bear in mind their own experience when we lift the screen covering B and increase the number of y's to add to A. However, since they are unable to understand the quantitative value of the covariations, they proceed qualitatively and simply admit that the greater the absolute increase in B, the greater the difference $y = B - A$ must be.

3. STAGE II

The second stage marks the beginning of regularity in the covariations between the x's and the y's with the initial reactions to the simple questions being correct. But when we ask the control questions, the answers obtained not only indicate the subject's lack of comprehension of the function, as will also be the case in stage III, but actually perturb the subject to the point of bringing him back to the level of stage I reactions. We will distinguish, in this respect, several types of reactions which we cannot yet classify as substages, because they are more or less contemporaneous. The first type (II A) is that of subjects who accept $y = 2$ for $x = 1$ when A and B are small, but who revert to $y = 1$ when we go on to large collections:

SOB (4; 10) takes $y = 1$ for $A = B = 3$ when we take away $x = 1$ and then immediately corrects himself to $y = 2$. He continues correctly: '*I take two each time.* – Why? – *Because you have 8* (which is correct and which he deduces by counting his tokens in A)'. We try the second control test, taking 1 token away and placing it in the reserve box instead of in B: he doesn't fall in the trap[2] and only takes $y = 1$, then begins again with

$y = 2$ when we transfer $x = 1$ from A to B. We then pass on to $A = B = 17$: 'I take 1 away ($x = 1$ in B). – *I replace 1* ($y = 1$). – Why 1? – *Because you took 1 away from me'*. He continues $y = 1$ until $A = 17$ and $B = 23$. We then return to small piles, $A = B = 2$: he immediately takes $y = 2$ for $x = 1$ until $A = B = 6$. But when we skip to $A = B = 10$, he starts taking $y = 1$ again!

We then try to increase x. After a trial on small piles where he reacts correctly ($y = 2$ for $x = 1$), we remove $x = 2$: '*I place 3.* – Why? – *Because you have too many'*.

Cases of this type are interesting from several points of view. Sob misses the proportionality $y = 4$ when we go from $x = 1$ to $x = 2$, which is normal, but he does add one unit (he takes 3 instead of 2). On the other hand, when we proceed to large collections ($A = B = 17$ or even 10) he does not increase the y's as do the subjects of level I B who believe that the difference increases with the increase in the absolute numbers of A or B. Instead he makes the opposite error which is an intelligent one in a way, i.e. judging that between two large collections A and B the same absolute difference is relatively smaller than the difference between two small ones (Weber's law!), he decreases it absolutely and passes from $y = 2$ to $y = 1$ for $x = 1$.

Type II B subjects, like the preceding ones, begin correctly but are perturbed when asked the second control question (in which the tokens are returned to the reserve box) which like Sob they solve correctly although they are not able to subsequently readapt to the main question:

MOR (6; 11) goes from $A = B = 2$ to $A = B = 9$ taking $y = 2$ for $x = 1$ each time. 'And if I do that (one token is taken from A and returned to the box)? – (He begins with $y = 2$ twice in a row, then changes his mind): *I place only one token since you put it there.*' But, when we come back to the transfer from A to B he stays with $y = 1$. We begin again at $A = B = 2$ and the series of reactions is the same, including the confusion manifested after the control question. When we finally increase the transfer from $x = 1$ to $x = 2$, Mor reacts like Sob with $y = 3$ instead of 4.

These cases show that even when the subject understands the function between the transfer x and the difference y for very small collections, he can conserve it only by induction, without comprehending its necessity. When we present him with another type of function in which he only has to take back from the reserve box what we put there (after it was taken away from him), and then bring him back to making transfers from A to B at the point where he had left off (for Mor 9 or 10 elements), he no longer knows which rule to follow and he stays with $y = 1$.

The third type (II C) is even more interesting because this time it is the simple request for an explanation (third control question) which confuses the subject and interrupts the inductive progression.

CUR (5; 4) each time takes $y = 2$ for $x = 1$ between $A = B = 2$ and $A = B = 11$. 'How many will you take? – *Two*. – And me? – *One*. – Why do you take 2 when I take 1? – *I don't know*. – Will we have the same that way? – *No*. – Who has more? – *Me*. – Why? – *Because I take 2*. – And now (we take $x = 1$)? – *I take 1*. – Do we have the same thing? – *No*.' We begin again at $A = B = 2$ and he reverts to $y = 1$ until the collection becomes too large whereupon he goes on to $y = 2$ but without being able to explain why.

CAR (5; 3) reacts in exactly the same way. When we start at $A = B = 6$, he stays with $y = 1$. When we go back to $A = B = 2$, he understands and begins to react correctly ($y = 2$). But at $A = B = 5$ when he is asked to explain, he again becomes lost and reverts to the system $y = 1$.

These cases show once again that as soon as the subjects go beyond very small collections, induction is possible without comprehension. The empirical character of this induction is here even more striking since it is enough for us to ask the subject the reason for his choice of the function $y = 2$ when $x = 1$ for him, on not finding it, to refuse to generalize and to revert to $y = 1$.

We could also distinguish a II D type (intermediate between stages II and III) present in subjects who although not disturbed by the control questions (they have succeeded in the first two questions and only fail when asked to give the reasons) revert to $y = 1$ for $x = 1$ when we push them to generalize for $x = 2, 3$ or n:

GIT (6; 10) answers everything correctly for $x = 1$ without however being able to give a reason. We move on to $x = 2$, and he again takes $y = 2$ (several times). When we go back to $x = 1$, he takes $y = 1$, although he had not made this mistake before.

In summary, stage II shows that a function which is understood for small collections of 1 to 3 tokens can be generalized inductively even while becoming less and less clearly understood. Thus on the operatory plane, this situation corresponds to what is, on the plane of physical objects, the beginning of the induction of laws as opposed to causal explanation.

4. STAGE III

This is the stage of regularity and inductive generalization which passes progressively from $2 y$ for $x = 1$ to $2 ny$ for $x = n > 1$. But since the passage is a very gradual one, we can distinguish the following steps. At level III A, the subject reacts correctly for $x = 1$ without being troubled by the control questions, but for $x > 1$ he reverts to $y = x + 1$ every time:

LAF (6; 1) for $A = B = 3$ with $x = 1$, takes $y = 2$: '*I guessed: I had 3 and you took 1*'. He continues likewise until $A = B = 7$: 'How many will you take? – *Two*. – And me? – *One*. – Why? – *To have as many as you*. – And if I put it there (reserve)? – *You have to replace it* (he takes 1 for *A*). – And now (*x* = 1 from *A* to *B*)? – *I take 2*' and so on until *A* = *B* = 13. 'Now look (*x* = 2). – *I take 3*. – Why 3? – *Because you took 2*. – And like that (*x* = 1 put back in the reserve)? – *One, since you put 1 there* (reserve), *you have to replace it here* (*A*). – And that (*x* = 1 from *A* to *B*)? – *Two, because you put 1 under your hand*. – And now (*x* = 3)? – *Four. You took 3, so to do the same thing I must take 4.*'

SUT (6; 8) and LOP (6; 11): same reactions, with a clear distinction as to the passage into the reserve box or into *B* and with generalization: 2 for 1, 3 for 2, 4 for 3, etc. as if *y* = *x* + 1 in all cases.

CHU (8; 7) reacts similarly with the sole explanation: '*3 because you took 2; it's to have the same thing as you*'.

The law *y* = 2*x* is thus well generalized and understood for *x* = 1 but given the lack of proportionality, it becomes *y* = *x* + 1 for *x* > 1 with a non-differentiation or confusion of the ordinal sequence of *x*'s and cardinal values of *y*'s.

Substage III B marks a slight progress in that the subjects arrive at *y* = 4 for *x* = 2, although by trial and error, and generalize *y* = 5 for *x* = 3 as if *y* = *x* + 2:

GIS (6; 8) reacts correctly for *x* = 1 from *A* = *B* = 4 to 10: '*I always take 2 and you always take 1*. – Why? – *You have more than me* (he means that when *B* has 1 more, *A* has 1 less). – But how? – *You took 1 away and I must replace 2*'. If the token is placed in the reserve: '*I take 1 because you placed it back in the box*'. For *x* = 2, he begins by taking *y* = 2: 'Why? – *You took 2. Oh!* (he adds 1, thus *y* = 4). – (Again *x* = 2). – *This time I will add 4*', and so on. But for *x* = 3: '*I take 5 because you took 3*. – How many did you have? – *14*. – And me? – *14*. – So? – *Oh! you've got 17*'. He adds 1 to 3 which gives *y* = 2*x*, but in the next test he reverts to *y* = 5 for *x* = 3.

We see that the subjects have come to understand the law for *x* = 1 and in certain cases to also understand it for *x* = 2 or even 3, but in general as soon as *x* > 1 they proceed only by empirical induction. At the III C level, the law *y* = 2*x* is understood right away for *x* = 2 as it is for *x* = 1, but starting with *x* = 3, the subjects revert to approximate inductions:

GEN (6; 9) starting with *A* = *B* = 4 to 14 consistently gives *y* = 2 for *x* = 1. We pass on to *x* = 2 and he immediately takes *y* = 4 justifying it this way: '*If you take 1, that makes 1 that I had and you have 1 more* (thus *y* = 2)' but if *x* = 2: '*You took 2. – So? – You have 2 more* ('*and me 2 less*' being understood). – And like that (*x* = 3)? – *It's hard! This time you need 1½ more* (he made a mistake in direction and divided instead of multiplying). *You need 5* (thus) *2 more*. – Why 5? – *You took 3, so I had 3 less; in order to have as many as you, I took back 5*. – Why 5? – ... – And that (*x* = 4). – *I take 6, you took 4 which gave me less, so I take back 6*'.

ROU (8; 6) likewise understands the reason for *y* = 2 for *x* = 1: '*You take 1* (in *B*), *you took one away from me* (in *A*), *thus to have the same amount as you, I take back 2*'. For *x* = 2 he begins with *y* = 3 but says immediately: '*No, 4, you have to count them as before* (thus +2 in *B* and −2 in *A*)' and he sticks to the rule thus understood. But for

$x = 3$ '*I take 5, that's 2 more.* – And if I take 4? – *Then it's 6'*. Nevertheless in the following he finds $y = 6$ for $x = 3$: '*Thus 6 to make the same amount* $(x + x)$'. But after returning to $x = 1$ and $x = 2$, he reverts to the solution of 4 for $x = 3$.

We can thus characterize this level as the passage from the ordinal solution $y = x + 1$ to a solution which although not yet proportional, except for $x = 2$, is in some respects hyperordinal, i.e. the increase in the y's is a function of the x's but with an additive increase due to the lack of a sufficient numerical evaluation of the A's and B's.

Finally, the fourth substage III D marks the discovery of the law $y = 2x$ but here again by an inductive process rather than by deductive construction as in stage IV:

JAN (6; 5) begins with $y = 1$ for $x = 1$ and after verification consistently stays with $y = 2$. When we pass on to $x = 2$ he immediately takes 4: '*Otherwise it's not the same thing*'. We come back to $x = 1$ then to $x = 3$: '*It's 6. When you take 3, I take 6 and that way it's the same*'. We skip to $A = B = 17$ and he maintains $y = 6$ for $x = 3$.

JAC (6; 6) starting with $x = 1$ says: '*I know how. I have to take 2 to have the same thing*'. For $x = 2$ he first of all places $y = 3$ then says '*I don't think that's correct*', whereupon he takes 4: 'Why? – *To make it exact*'. For $x = 3$, he immediately places $y = 6$ and continues to do so three times in a row. 'And like that $(x = 4)$? – *Wow! 4, I've got to take 8.* – Why? – *To make it exact.* – And that $(x = 5)$? – *I have to take 10.*'

LAC (7; 10) right away takes $y = 2$ and 4 for $x = 1$ and 2. For $x = 3$ he tries to count what must be in the hidden pile B and takes $y = 6$. For $x = 4$ he right away takes 8: 'Did you count? – *No, 4 + 4 = 8, I know it by heart*'.

CAR (8; 4) begins with $y = 1$ for $x = 1$ but observes the error when B is uncovered and subsequently takes only $y = 2$. For $x = 2$: '*I take 4, because you took 2: so I take 4 instead of 2*'. We start again at $A = B = 6$ with $x = 1$, then $x = 2$. He takes 4 '*because you put 2 and I took double.* – And if I take 4? – *I put back 8.* – Why? – *I took double. I must not take 4 because then I would have less.* – And that $(x = 3)$. – *Six*'. But it is doubtful that he really understands the reason because up to the very end he takes $y = 2$ whenever we take away a token and put it back in the reserve box.

We thus see how the exact relationships understood for the small collections A and B and for the transfers x of 1 or 2 lead to an inductive generalization which in turn leads to the attainment of the function $y = 2x$ for any value of x (let us not exaggerate: only up to $x = 4$ or 5). However this induction still lacks necessity and while the law discovered is in a sense recurrent, it lacks constructive deduction. This is made evident by the one subject who explicity mentioned 'double' but who still failed one of the control tests.

5. STAGE IV AND CONCLUSIONS

In the final stage we find 'complete induction', a term sometimes used by Poincaré to designate recurrent reasoning, i.e. a constructive deduction which of necessity generalizes for the remaining numbers what was demonstrated for the first ones. We can again distinguish two substages: IV A in which the subjects proceed by concrete recurrence and IV B in which they proceed by formal explanation. The following are examples of IV A:

FER (8; 1) right away takes $y = 2$ for $x = 1$ and $A = B = 4$ '*because you take one more away from me* (than A had to begin with) *and if I take only 1 that will make 1 less than you*'. For $x = 2$: '*4 because you take 2. Without that I have 2 less than you*'. For $x = 3$, $y = 6$ and $x = 4$: '*8, otherwise I will have 4 less than you*'. And 4 in the reserve box: '*I take 4: when you put 4 under your hand you have more so I need to take more*'. For $x = 5$, $y = $ '*10 because you took 5*'.

DUV (8; 2) reacts like Fer but expresses himself more clearly: '*When you take 1 token, it increases yours by 1 and decreases mine by 1 token*'. For $x = 2$, $y = 4$ because '*when you take 2 tokens, it increases yours by 2 and decreases mine by 2 tokens*'. And so on.

DEV (8; 6): '*You already have 1* (when it was taken away A became 9), *and you took 1, that gives you 2 more, so I take 2*. For $x = 2$, $y = 4$ '*because you took 2, and I therefore had 4 less*'. For $x = 3$, same reasoning '*then I must take double that*', etc.

And here are some cases of level IV B:

KIM (9; 3) begins like Duv and Dev for $x = 1$. From $x = 2$ onwards, he generalizes: '*I take 4 to make the same amount: we always put double the amount*'. Same for $x = 3, 5$, etc.

ALB (9 years): '*I have to take double, one* (one-half) *to have the same as before* (in A) *and the other for what you took* (in B). *I must always take double*'.

MAT (10; 9): '*I must take what you took away and what you have extra*'. If 2 tokens taken from A are placed in the reserve box, he says: '*I take 2, because you don't have any extra*'.

CLA (11; 7): '*It's double each time: you have* (for $x = 4$) *4 more and me 4 less*'.

We have thus arrived at the end of a developmental process whose general framework seems clear. Although the subject does not doubt at the outset that a transfer of n tokens from one collection to another will result in a difference of $2n$ between them, he is at first only able to ascertain this relationship in some cases without being able to apply it to others. He then gradually begins to generalize until he subsequently discovers the complete generality by recurrent induction after which he finally (IV B) finds the reason.

From the standpoint of functions this raises two problems: that of the reason for the initial lack of comprehension (which concerns 'constitutive' functions), and that of the progressive integration of the function into an operatory structure which provides the reason (which concerns the 'constituted' functions).

I. – The initial lack of comprehension and the very strong tendency to believe that a transfer of n tokens results in a difference n rather than $2n$ raises the general problem of covariation. In contrast to a non-functional relation which is only the result of a comparison, a function always expresses a dependence, and in fact, in this particular case, none of our subjects doubted that the difference y to be calculated 'depended' on the transfer x which induced it. Yet the dependences between y and x in the function $y = f(x)$ can be presented in one of two ways depending on whether the action, whose scheme the function expresses, tends to transform its object or simply find it again, reproduce it, or replace it by an equivalent. In the latter cases where the function takes on the form of a simple application, the dependence is expressed as a 'coproperty' while in the case where the function is a transformation, it is expressed as a 'covariation'. We can thus admit that in stage I of the present reactions, young subjects who assume $y = x$ instead of $2x$ ignore the covariation between the two collections Y and Y' and orient themselves towards a particular coproperty. Why does this happen?

The reason is not, as we might tend to believe, solely based on the fact that one of the collections is always visible while the other is periodically hidden. In fact, as we will see in the following chapter, the reaction is identical even when both collections are always visible and the task is one of setting up an inequality between them. (In that experiment, young subjects expect the difference to be n when they transfer n tokens and so are very much surprised when they always get $2n$.) Rather it seems to us that the explanation derives from the fact that in the relations between two systems, it is easier to find a one-way action than a double movement with a return action. In the domain of causality, this is what explains the fact that the notions of reaction, interaction, circular causality and feedback have taken so much longer to become accepted than linear causality, in spite of the fact that the latter occurs rarely if ever. In the operatory domain the present example is analogous: given two systems constituted by equal collections of tokens, the subject either focuses on the one which loses n elements, neglecting the increase in the other and taking back n elements to reestablish equality (as is the case in this chapter), or he focuses on the one which gains n elements and forgets the correlative decrease in the first (as we will see in the following chapter). In other words, the subject acts (and consequently reasons) in one direction and looks for the equality between what he has lost and what he must recover instead of thinking immediately in terms of schemes of covariations or interdependences: thus the primacy of the coproperty $y = x$.

II. — The passage from the coproperty to the comprehension and deductive rationalization of the function $y = 2x$ is arduous because the subjects must pass from the simple function to its explanation which alone can confer on it the status of necessity. This function is first of all observed as a fact and used as such (stage II with generalization in III), but only as an induction comparable to physical induction except that it bears on the results of the action of a subject. It is then generalized for all numerical sets (level IV A), but thereby only acquires the status of being 'always true' which is not as yet the same as internal necessity. To attain the latter, one must both distinguish between and coordinate as the subject Mat does, 'those which you took from me and those which you have extra', even though they are the same tokens. In other words the same number n intervenes in two forms, n subtracted from system I and n' added to system II, thus $n + n' = 2n$ which is the difference y, even when we have $n \equiv n'$ within the transfer x.[3] In this case the function $y = 2x$ is no longer sufficient unto itself and is no longer merely observed, i.e. it is integrated into an operatory system whose covariations it expresses, and it is this system as such which confers necessity to it since by itself it is only a law. (In this particular case a law of coordination of actions, but one which could express a physical law in other situations, as we noted in the introduction to this chapter.) We thus have an evolution from a constitutive function to a constituted function in the sense of the passage from a simple function to an operatory composition, as, in other domains, we attend the passage from functional laws to causality.

NOTES

[1] By A. Szeminska and J. Piaget.
[2] He thus understands that he only has to take back from the box the token which we have just taken from him and put there.
[3] $n \equiv n'$ expresses the identity of the two terms and not only their equality.

CHAPTER 6

THE COMPOSITION OF DIFFERENCES:
UNEQUAL PARTITIONS[1]

The preceding chapter dealt with an elementary function which was causal as well as operatory and whose progressive operatory treatment we analyzed, observing that the inductive discovery of the underlying law is possible from about age 7. It might now be interesting to examine a somewhat more complex function which would encompass the previously studied one. In other words, we will study the operatory composition of functions. The following task is useful in this respect: given a set with a numerical value of x divided into two unequal parts whose difference is x', find the values y and y' for the two parts.[2] For example, if 30 tokens are to be divided between two boys such that one is to have 6 more than the other, the object is to find that one should receive 18 and the other 12. It's clear that if the parts were equal (15 and 15), it would be necessary for one to take 3 tokens from the other in order for the difference 3 + 3 to equal 6, which brings us back to the problem presented in the preceding chapter.

In fact, the two possible methods for finding the values y and y' of the two parts are (if x is their sum and x' their difference, x and x' being the variables given and y or y' the variables to be found):

(1) $$y = \frac{x}{2} + \frac{x'}{2} = \frac{x + x'}{2} \quad \text{and} \quad y' = \frac{x}{2} - \frac{x'}{2} = \frac{x - x'}{2}$$

or

(2) $$y = \frac{x - x'}{2} + x' \quad \text{and} \quad y' = \frac{x - x'}{2}$$

We note that in method (1) the difference x' between the two collections must be divided by two, which is the reciprocal of the task in chapter 5. Furthermore, even though this composition seems complicated, it in fact only comprises additions, subtractions, and halving while the problem in the preceding chapter involved, besides addition and subtraction, only duplications. Thus the operations involved are the same and the stages observed correspond closely to the preceding ones, although with a certain *décalage*.

1. TECHNIQUE AND GENERAL RESULTS

Since our objective was to uncover the operatory mechanisms of the composition of these functions y or $y' = f(x,x')$, we presented the problem on the basis of concrete examples, although we first asked for a deductive solution based on an oral or written calculation before passing on to the actual manipulation of the objects and to induction based on empirical observation.

We began by placing a set of 30 tokens before the child and by asking him a question in the form of a story. A mother has two sons, A and B to whom she offers these tokens but who for some reason (B had more previously, etc.) wants A to have 6 more than B. The task is thus to find out how much A and B will each have. The subject is allowed to look at the tokens as much as he wants but is not allowed to handle them and the solution must be obtained either orally or in writing. If the subject is not able to work with these numbers, we go on to a simpler problem where he must divide 10 tokens with a difference of 4, here again with oral or written calculations. Once the deductive capabilities of the subject have been fully exhausted, but only if he has still not succeeded, he is asked to solve the same problem concretely, by manipulating the tokens.
    If the subject still has difficulty with the concrete method, we go on to various elementary allotments, both in order to judge the level of the subject as well as to facilitate his subsequent solutions: (a) 'How many more are there' from 1 to 3 tokens? (b) 'How many must you add here (1) to have the same as "there" (3)?' (c) 'How many must you take away from here (3) to put there (1) if we want both boys to have the same?' For manifest reasons this question (c) is the same as the one asked in the last chapter.

We can distinguish the following four stages which generally correspond to those of Chapter 5, even though we only questioned children age 7 to 13 since we expected greater difficulty. However, since stage I is still found at age 7, this is of no real importance. This stage I is characterized by two reactions the second of which is identical to that of stage I in Chapter 5, i.e. either the subject considers the totality as formed by x (the total indicated) $+ x'$ (the difference given) and then he simply divides $x + x'$ into 2; or he divides x into two halves and adds to one the difference x'. This in fact comes to be the same as introducing a difference of $2x'$. In stage II, the subject starts with analogous solutions, but only achieves some success by empirical trial and error, without however generalizing this for the subsequent tests. Stage III again (cf. Chapter 5) involves the progressive inductive discovery of the function. Finally in stage IV, the subjects arrive at the solution by deduction prior to manipulation, on the basis of concrete operations in IV A and purely by calculation in IV B.

2. STAGE I

Here are examples of these initial non-coordinations:

RIT (7; 1) does not succeed in any calculation for 30 tokens. For $x = 10$ tokens with a difference x' of 4 he finds 2 and 7, then 5 and 9 which does give $x' = 4$, but for a total of 14. We then invite him to handle the tokens whereupon he divides 10 into two halves and gives $5 + 4 = 9$ (to A) and $5 - 4 = 1$ (to B). 'How many more does (A) have? – (He counts) 8. – And how many do you need? – 4 (he returns the 4 to B, thus $5 = 5$). – But make him have 4 more. – *I give 1 to* (*A*) *which makes 6 for* (*A*) *and 4 for* (*B*). – How many more does (A) have? – *One* (he passes 2 more from B to A, thus 8 and 2). – How many more does (A) have? – 4. – How many tokens does he have? – 8. – And (B)? – 2. – Well? – (He returns 4 to B then equalizes them at $5 = 5$ and gives 4 tokens to A!)'.

He is able on the other hand to divide 3 with $x' = 1$, by giving 2 to A and 1 to B. For 5 tokens with $x' = 3$, he gives 3 to A and 2 to B and since the difference x' is then only 1, he transfers the 2 from B to A which makes 5 and 0, then takes the 2 from A to B which again gives 3 and 2, etc. He just can't seem to get it. For 4 tokens with a difference x' of 2, he starts out by giving 2 to A and 2 to B, then transfers the 2 from B to A which makes 4 to 0 and then he starts over again without finding a better solution. We show him the solution and he reproduces 3 and 1. We again ask him to divide 5 with $x = 3$ and as before, he gives 3 to A and 2 to B, then transfers the 2, which leaves 5 and 0 before finally discovering that in taking back 1 token from A to B, he leaves 4 and 1. We then propose a partition of 7 giving 3 more to A: he puts aside 3 tokens, and of the 4 remaining ones gives 1 to B and 3 to A to which he adds the other three, thus 1 and 6; then he starts over again and finally gives 4 to A, 3 to B and then, in going from this $x' = 1$ to $x' = 3$, he transfers 2 tokens from B to A: thus 6 and 1 again. He gives up.

COB (7; 6) calculated in writing for 30 tokens with a difference of 6. He gives half to each plus 6 to A, which gives $x = 36$ instead of thirty. He then manipulates the tokens but still can't get it to work. For $x = 10$ and $x' = 4$ he concludes that A will have 14 and B will have 10, then divides the 10 tokens into 5 and 5 and transfers 4 such that $A = 9$ and $B = 1$: '*That won't work*', he removes 5 from A leaving 4 and 6: '*The other has 2 more*', so he moves these 2 which gives him 6 and 4. We then ask him to use $x = 4$ with $x' = 2$ and he correctly finds 3 and 1. Next $x = 5$ and $x' = 3$: '*One has 4, the other has 1, that makes 3 more*'. After these successes we ask him for a set of 8 with $x' = 2$. He divides this into $4 + 4$ and displaces 2 which gives 6 and 2 but he just can't seem to get it.

ZER (8; 2) succeeds for $x = 3$ and $x' = 1$ and for $x = 5$ and $x' = 1$, but by first having given 4 and 1 then 3 and 2. For $x = 4$ and $x' = 2$ he gives 2 and 2, and 4 and 0, 2 and 2, 4 and 0 and gives up because he doesn't understand that when he transfers 2 the difference is 4. We show him 3 and 1 and then ask him to divide 10 with a difference of 4: he gives 6 and 4, transfers 4 and obtains 2 and 8, does the opposite and comes back to 6 and 4, etc. He gives up.

LER (9; 4), in spite of his age, does no better. For $x = 30$ and $x' = 6$ he calculates 14 and 10 and neglects the rest. For $x = 10$ and $x' = 4$ with manipulation, he starts with 5 and 5, transfers 4 and finds 9 and 1: '*Oh! I would never have thought that there would be 1 left!*' He starts out again at 5 and 5 then returns to 9 and 1. He then tries transfers by 1 or 2, thus arriving at 4 and 6, 2 and 8, then again at 4 and 6, after which he gives up.

These facts could not be any clearer. When the subject calculates in his head, he forgets the whole, like Cob who ends up with 36 tokens when there are only 30, or like Ler who calculates with only 24 tokens, forgetting about the rest. When manipulation is permitted, the subject proceeds by transfers but is not able to understand that when n tokens are moved he necessarily obtains a difference of $x' = 2n$ rather than n. This is the exact equivalent of the stage I reactions to the experiment in the preceding chapter, but it is much more significant here since instead of having one hidden by a screen, both collections $A(y)$ and $A'(y')$ are visible. Only the small sets comprising 3 to 5 tokens allow correct solutions which are perceptual rather than reasoned (nor even generalized: cf. Zer for $x = 4$ and $x' = 2$), and thus are not generalized for larger sets of 8 or 10.

3. STAGE II

Corresponding as it does to stage II of the preceding chapter, this second stage is characterized by the empirical discovery of some partial solutions without as yet any generalization of the law:

SUD (8; 6) fails when asked to mentally divide the 30 tokens. For $x = 10$ and $x' = 4$, he starts with 5 and 5, moves 4 and finds 9 and 1, then 8 and 2, and 6 and 4 without understanding what he is doing. He succeeds in dividing 5 with the difference $x' = 1$. For $x = 6$ and $x' = 2$ he divides it into 3 and 3 and displaces 2 thus obtaining 5 and 1. After looking at the two collections and moving 1 he finds 4 and 2: '2, *that's right*'. But he is not able to generalize the method for 7 with $x' = 5$: he tries 3 and 4, 2 and 5, 7 and 0 and then finally gives up.

ZIE (8; 6) also fails with the 30 tokens. For 10 with $x' = 4$, he also begins with 5 and 5, transfers 4 which gives 9 and 1, 6 and 4 and finally 8 and 2. He then displaces 2 several times, obtaining 6 and 4, 8 and 2, etc., and ends up by displacing 1 instead of 2, thus obtaining 7 and 3: '*There, that makes 4 more*'. In the following tests (11 with $x' = 5$; 13 with $x' = 5$ and 15 with $x' = 7$), he begins each time with transfers of 2 starting with 5 and 6, etc. and then, only after vacillating between + and −, finally arrives at the last transfer of 1 which assures the solution.

SEN (8; 6) tries to divide the 30 tokens with a difference of 6 by first proposing 9 and 21, then writing $30 - 6 = 24$ and then proposing 24 for B and $24 + 6$ for A. For 10 tokens with $x' = 4$, he begins with 6 and 4, puts them back in a pile and divides them into 8 and 2: '*That makes 6. Since there must be 4 more, I take away 2 and add 4, which makes 10 for* (A) *and none for* (B)'. We pass on to 5 with $x' = 2$ which he succeeds in doing and then we come back to 10 with $x' = 4$: he starts over with 8 and 2 then 6 and 4, transfers 2 several times and seems surprised, then understands: '*If I remove 2, then that makes 4* (the difference). *If I take 1* (from 4) *that makes 4 more* (7 and 3)'. He bears this empirical discovery in mind when dividing 13 with $x' = 3$ (3 and 10, then 5 and 8) but fails to apply it when dividing 18 with $x' = 6$: he tries 5 and 8, then 7 and 11, etc. without finding the solution.

We thus see how even in cases where there is local comprehension (Sen) success by empirical adjustments does not yet lead to generalizations. There are nonetheless some partial generalizations (like when Sen passes from

$x = 10$ to $x = 13$) but since these arise gradually step by step, we could multiply the substages of this stage as we did in stage II of the preceding chapter.

4. STAGE III

This stage marks the inductive discovery of the law, i.e. once a function has been empirically isolated in one or two cases, it becomes generalized for all cases even though the common form is not deduced:

OPA (7; 5) fails to divide the 30 tokens. For $x = 10$ and $x' = 4$, he starts with 5 and 5 and by a transfer arrives at 9 and 1, etc. He then starts over with 5 and 5, taking 2 from each, which constitutes the difference of 4 and places these 4 tokens aside summarizing the situation as follows: '*For (A) to have 4 more, you must give 3 to (A), 3 to the other and 4 more to that one (A)*'. This is in fact function (2) indicated at the beginning of this chapter. For $x = 11$ and $x' = 5$ he follows the same procedure: '*They each have 3 with that one having 5 more*'. For $x = 16$ and $x' = 5$ (insoluble problem except when $y = 10\frac{1}{2}$ and $y' = 5\frac{1}{2}$), Opa does the calculation in his head and finds that '*I can't do it: 5 + 5 doesn't work, 4 and 6 don't work, etc*'. – 'Would it work with the tokens? – *I think so.* – Is it the fault of the numbers 16 and 5 or your calculations? – *My calculations.* – But your calculation is correct!' He nevertheless takes the tokens and tries, then states: '*It's because of the number, because it's an odd number*'. We propose $x = 11$ and $x' = 7$; he removes 7 and divides the other 4 by two: '*Each has 2 and (A) 7 more.* – Does it work then? – *Yes, 2 and 9.* – Why doesn't 16 work with a difference of 5? – *Because when we take away 5, we have 11 tokens left which can't be divided*'.

PIT (9; 2) fails in the mental division of the tokens (15 plus 21, then 9 and 15 and finally 24 and 6). When allowed to manipulate them, he starts with 15 and 15 and after transferring 6 arrives at 9 and 21. He then starts over again and gives '*3 to each*', distributing them 3 by 3 up to 12 for A and 12 for B and giving the 6 remaining ones to A. For $x = 10$ and $x' = 4$ he starts with 5 and 5, passes on to 9 and 1, then begins to distribute the difference by 2's which immediately gives 7 and 3. For $x = 11$ and $x' = 5$, he first of all observes that we can't divide halves and by applying his preceding method, finds: '*They must each have 3 and that one 5 more*', thus 3 and 8. For $x = 15$ and $x' = 7$: '*I want to put 7 more* (he sets them aside) *then there are 8 left: I then give 4 to each and then 7 more to that one*'.

DUL (10; 1) also fails in dividing the 30 tokens and begins for $x = 10$ and $x' = 4$ with the solutions 9 and 1, etc., and finally finds: '*Each will have 3 and (A) 4 more*'. For $x = 16$ and $x' = 6$, he divides it into 8 and 8 then transfers 3: 'Why 3? – *It's half of 6.* – Why half? – *It's 3 from one and 3 from the other.* – And with 12 and 4 more to (A)? – *Then 6 + 4 = 10. Oh no! That's wrong: it's 6 + 2 = 8 and 6 – 2 = 4*'. He succeeds likewise in dividing 29 tokens with $x' = 7$, but then for 17 with $x' = 6$, he observes that '*it doesn't work, those are odd numbers*'.

As we can see, these reactions which can generally be observed towards 8–9 years of age (Opa is no doubt advanced) are significant. The subjects begin by trial and error but quickly understand during these attempts that a transfer of n tokens from one pile to the other results in a difference x' of $2n$, which as we recall is the discovery made in stage III of the preceding chapter. Once this has been grasped, certain subjects (like Dul) start with half of the whole (or $x/2$), then divide the difference x' into two equal parts leaving one with A and taking the other away from B to give to A. Other subjects like Opa and Pit skip this stage and look immediately for half of $x - x'$: thus for y (A) the sum $(x - x')/2 + x'$ and for y' (B) the value $(x - x')/2$. This inductively discovered method is then generalized as a method, not yet as a deductive demonstration, to the degree that the child is able to understand the impossibility of a solution when $(x - x')$ is an odd number (Opa and Dul). It should be pointed out that Opa does not as yet sufficiently believe in deduction to be able to trust his calculations, but rather needs concrete manipulations in order to be convinced.

5. STAGE IV AND CONCLUSIONS

This last stage is characterized by the deduction of solutions through calculation. Before level IV A, subjects are able to make mental generalizations, but they still need a certain amount of manipulation, or at the very least, some written trial and error attempts:

BOR (9; 5) mentally divides the 30 tokens into 21 and 9 to make $x' = 6$ and is not able to find a solution before being allowed to manipulate them. As soon as he can do so, he moves 3 from A to B: 'Why 3? – *To make 6*'. For 18 with $x' = 4$ (without tokens): '*9 and 9 makes 18. It's 9 each and 9 + 2 for (A)*. – And the other? – *9 minus 2 is 7*'. The same thing occurs for 21 and $x' = 5$ where he finds (by writing) 13 and 8 adding $10\frac{1}{2}$ and $2\frac{1}{2}$, etc.

CHA (10; 2) mentally divides 30 (with $x' = 6$) into 15 and 15, adds 6 and finds 21 and 9: '*12 (difference) is too much: you need 3. One will have 12 and the other 18*. – Why 3? – *Because it won't work with 6, you need half*. – And 48 with a difference of 14? – *It's 24 and 7 more: that makes 17 and 31*. – And 29 with a difference of 7? – (He divides it into 14 and 15 displaces 3): *It's 18 and 11*'.

BAT (11; 1) reacts like Cha for 30 with $x' = 6$. For 48 and $x' = 14$: '*Half of 48 is 24 and I add 7 more because 14 is too much: that makes 24 and 7*'.

DUC (11; 5) fails to mentally divide 30 with $x' = 6$. When given the tokens, he sets aside 6 and divides 24 into 12 and 12: '*It's 12 and 18*'. For the other problems, he applies the same method but does so mentally.

At level IV B, the problems are solved mentally with a minimum of trial and error:

ZUM (11; 4) for 30 and $x' = 6$: '*First of all 15 each and 6 more to (A) makes 21. No, I made a mistake. I must add 3 and take away 3 which makes a difference of 6: 12 and 18*. – And 25 with a difference of 7? – *I give 12 to (B) and 13 to (A); and 3 more to (A) and 3 less to (B): 9 and 16*. – And 29 with a difference of 11? – *I take half, 15 and 14, then I take away 5 from 14 and add them to 15: that makes 9 and 20*. – Find another method. – *We take away 11 from 29 and divide by 2 which gives us 9 and then we add 11 to the other*'. For $x = 28$ and $x' = 7$, he finds $17\frac{1}{2}$ and $10\frac{1}{2}$ and concludes that this cannot be applied to the tokens because 7 is an odd number.

DUR (11; 8) right away solves the problem of 30 with $x' = 6$: '*It's 18 and 12.* – How did you do it? – *15 and 15 and, if there is a difference of 6, one has 15 + 3 and the other 15 – 3.* – Why not ± 6? – *That would be a difference of 12.* – What about 33 with a difference of 9? Can you find another method? – *I take away 9 and divide by 2* (thus 12), *which makes 21 and 12'.*

FEL (12; 2) for 44 with $x' = 14$: '*44 divided by 2 is 22. To one we give 22 + 7 = 29 and to the other 22 – 7 = 15.* – And 43 with a difference of 9? – *That's more difficult. $43 \div 2 = 21\frac{1}{2}$ and 9 more, no $4\frac{1}{2}$ to (A): that makes 26 and 17'.* For 47 and $x' = 11$ he finds 29 and 18 the same way. 'Find another way. – *Take away 11 and divide the rest into 2 and add 11 to (A)'.*

We thus see that at level IV A which corresponds to the level of concrete operations, the subject's reasoning can again be characterized as a mental experiment with successive corrections as a function of the results obtained. (Cf. Cha: 'Because it won't work with 6, you need half'.) This does not mean that the mental experiment both constitutes and explains the reasoning as Rignano thought because it is directed, as is the physical experiment, by operations which are not drawn from objects; it does however indicate that at this level, operations consist of mental manipulations analogous to real ones. At level IV B, where the operations become hypothetico-deductive, they permit an anticipatory and mobile deduction and we cannot help but admire the speed with which subjects such as Dur and Fel change the method used upon request and take turns in using the solutions (1) and (2) presented in the introduction to this chapter.

As regards the general significance of these stages from the standpoint of functions, we see that the successive reactions to this problem (which is the reciprocal of the problem of equalization dealt with in the preceding chapter) occur in the same order with a fairly weak chronological *décalage*. We could have expected the latter to be even greater given the cases observed by L. Johannot[3] and above all given the fact that our questions were not only directed at finding the necessary transfer from collections A to B, the difference between them being given, but furthermore at finding the numerical value of these A and B collections, given only their sum and the difference to be conserved. The relatively precocious character of the solutions found (stage III and IV from ages 8–9 and 10–11) and the mobility of the methods of stage IV shows that as soon as a constituted function has been integrated into an operatory system (and therein lies its definition) it is concomitant with its composition or compositions and thus makes new ones possible. At this level, functions which previously expressed the covariation between any given variable or even the coproperty between the terms of any given action, are now no more than the dependences between operatory transformations.

NOTES

[1] By A. Szeminska and J. Piaget.

[2] The partition must naturally be exhaustive.

[3] Johannot, L., *Le raisonnement mathématique de l'adolescent*, Delachaux, 1947, p. 23 ff. The problem studied by this author (If we have the same sum of money and I give you 23 francs, how many more francs would you have than I?) is in effect resolved only starting from age 12–13. But its difficulty no doubt stems in great part from the fact that the starting values (the common sum) are not specified, which contributes to making the subject forget the subtraction from A and the addition to B at the time of the transfer.

CHAPTER 7

AN EXAMPLE OF THE COMPOSITION OF THE VARIATIONS OF VARIATIONS[1]

The situation which served as an example for this analysis of the composition of functions comprises, on the one hand, links which are clearly physical|or causal (the draining of a liquid with levels determined by the different shapes of the containers) as well as the operations properly so-called of serial correspondences (the multiplicative grouping of transitive asymmetric relations). This provides us with an opportunity to study the relationships between the causal and the operatory aspects of functions at the same time as the problem of their composition, both of these questions naturally being indivisible.

1. TECHNIQUE AND GENERAL RESULTS

The materials used consist of three jars (Figure 10). The first A is cylindrical in shape and is placed over jars B or C into which it drains through a cock, in equal quantities corresponding (on the cylindrical part of the glass) to heights which we shall call HA_4, $HA_3 \ldots HA_1$. The equal intervals DHA are observed by the subject between the markings on the glass. Jar B is also made of glass and has two large parallel triangular sides joined by two small sides and by a base 1 cm square (this is to avoid the three dimensional problem which the use of a pyramid would have raised). We will call the successive levels HB_1 to HB_5 and DHB_{1-4} the decreasing height intervals occupied by levels $HA_4 - HA_1$ from jar A; and we will call the increasing widths of the levels of the

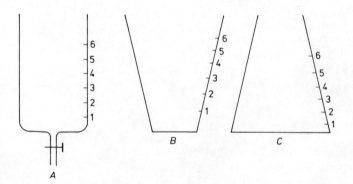

Fig. 10

water in B, WB_1 to WB_5 and DWB_{1-4} the decreasing intervals between these widths. A third jar C has the same shape as B but is inverted. We will call DHC_1 to DHC_4 the decreasing levels occupied by the equal quantities of water corresponding to DHA_4 ... DHA_1; WC_1 to WC_5 are the decreasing widths of the levels of water and DWC_1 to DWC_4 their increasing differences.

Given the above, the procedure adopted is as follows. We begin by showing the complete drainage of the water from jar A into jar B without any reference to possible interruptions, to the height intervals DHB nor to the widths WB_n. After this we announce that we will proceed by stages and thus ask the subject to anticipate the intervals DHB_1 to DHB_4. For this purpose, we give the child a rectangular strip of paper which can be placed vertically against jar B and on which the child simply indicates the different superimposed levels by marking them with a crayon. We then ask him to also predict the widths WB_1 to WB_4 and to help him we give him a new strip of paper, which this time is long and is presented horizontally on the table so that starting from the same point of origin he can indicate the successive widths by making simple marks thus enabling him to judge them graphically thanks to the class inclusion of the differences DWB. Next we ask the subject to indicate, if he can, how he went about predicting these two things and then we provide him with the means to verify his anticipations. We drain the liquid as he watches so that he can accordingly correct his anticipations regarding the height intervals DHB_1 to DHB_4, but without yet talking about his predictions on the widths WB. Once these corrections have been made, we ask the subject to correct his predictions as regards the widths before proceeding to a new draining. Then we again drain the water making the subject control the widths WB and the differences DWB which had been predicted on the drawings. Lastly we ask the subject to explain the succession of the intervals DHB as a function of the intervals DHA and the succession of the widths WB as a function of the heights HB, etc.

We then pass on to jar C and the procedure followed is the same. But once this new part of the questioning is ended, we ask for an explanation regarding the differences between the DHC's and DHB's and between the DWC's and the DWB's. In the last part of the experiment, we ask the subject to indicate which of the jars, B or C, is used when we show him, by means of a screen with a vertical slit, the variations in the height of the intervals without his being able to see the widths.

The observations gathered on the subjects aged 5 to 11 and on some adolescents aged 13–14 make it possible to distinguish 3 stages. During the first stage, which lasts up to age 7 or 8, the only functions used reveal immediate one-to-one correspondences, a complete isomorphism being assumed between the events in A, B, and C, including their directions. The intervals between the levels of water are thought to succeed each other from top to bottom in B and C as in A (except in the intermediate cases between stages I and II where they are equal to each other as well as to the DHA's; the same applies to the differences between the widths DWB and DWC which are anticipated as equal). On the other hand, once these general equivalences have been anticipated they are corrected when confronted with factual observations, although the latter do not in any way influence the subsequent predictions and consequently do not result in a comprehension of the functions involved. By contrast during the second stage (ages 7 to 10 on the average), the subject comes to predict that the direction of the successive levels in jars *B* and *C* which are being filled will be the opposite of that in jar *A* which is being drained. But the child does not yet anticipate the variations in the intervals of height or width. In contrast to stage I the confirmations which follow result in the construction of correct functions and in the immediate beginning of composition, but with some difficulties which may be due, among other reasons, to the graphic means used to mark the widths. Finally, during the third stage (age 11 and above), functions are correctly anticipated and composed in an overall morphism which encompasses all of the covariations in play.

2. STAGE I

In the experiments involving the conservation of the quantities of liquids transferred into containers of different shapes (for example from a lower and wider jar I to a higher and thinner jar II), we observe that if we hide the second container behind the screen, younger subjects expect a sort of integral conservation, i.e. they expect the quantity as well as the height of the column to remain the same. However, this is only a pseudo-conservation since it only applies to the levels and, when we ask the subjects to pour 'fairly' the same quantity to drink into the empty glasses I and II (similar to those which we have just described), they pour two quantities which are equal in height without taking into account the width of the glasses.[2] In fact when we return to the initial experiment, this time lifting the screen on II so that the transfer is clearly visible, these children immediately conclude that the quantity has not been conserved since the levels have changed. In contrast, when glass II is hidden, certain 5 year-old subjects already expect the level of water to be higher in II than in I since glass II is thinner. This however is only a prediction (lawful) based on previous observations with no comprehension of compensations, since on the basis of this anticipated difference in the levels these subjects still conclude that the quantities will not be conserved.

It was useful to recall the preceding facts in order to understand stage I and the passage to stage II of the present experiment and to show the need for a careful distinction in the analysis of the functions being constructed between the plane of anticipations (indices of the original levels of comprehension), that of observations and that of the final post-observational comprehension.

The anticipations proper to stage I reveal the most general and, no doubt, the most primitive of functions (as do the pseudo-conservations just discussed), i.e. an assimilation of the structures assumed to exist in glasses B and C to the structure observed in glass A. In other words, what we have here is a 'transport of structures'[3] where the subject proceeds by simple applications (characterized by equality of intervals, difficulty of inversions, etc.). This attitude is so tenacious that it leads to the identification of the direction in which B or C is filled from bottom to top) and the direction in which A is drained (from top to bottom) in spite of the observation made at the beginning of the presentation. It is so tenacious that even after detailed observations, the subject often returns to the total isomorphism of the levels in A, B, and C:

PHI (6; 0) begins by drawing unequal and incorrect intervals in B from top to bottom (the largest interval at the top and the smallest at the bottom), no doubt because he is thinking of the absolute amount of liquid in A which decreases progressively (unless he is influenced by the decreasing width of B, from top to bottom, although what follows shows that he ignores this factor entirely). When we ask him if B is filled starting from the bottom or from the top (before we have been able to question him as to the significance of these initial inequalities), Phi exclaims: *'Oh! I made a mistake, because it's the same here* (he shows the top of B and corrects his initial drawing by marking equal intervals from bottom to top). – How do you know that you must do the same? – *Because it's the same here* (he shows the intervals in A). – But there (in B: we show the bottom and the top), are they all the same? – *You must do the same thing*. – But here (we show the top of B) is it bigger, smaller or the same as here (the bottom of B)? – *It's the same*'. As regards the widths WB, Phi begins by indicating them arbitrarily, then spontaneously corrects himself and makes equal marks for the DWB intervals. We pass on to the verifications and Phi observes the inequality of the decreasing DHB intervals: 'Where is the largest? – *That one*. – Why does it become smaller and smaller. – *There's no space*. – Why? – *Because we made it too high here* (= because we put too much water at the bottom in the beginning). – But why when the syrup rises is the little bit which we add (we show the intervals on A) still so small? – *There is too much water here* (in the bottom)'. As regards the widths, if they are shorter towards the bottom it is *'because there was not enough space, that's why it becomes smaller'*. But these observations with explanations which among others thus reveal the non-conservation of the drained quantities do not in any way influence the predictions regarding C except immediately after a suggestion has been made. Phi gives again, in effect, intervals equal in height as well as width. At the time of verification, he recognizes the inequalities DHC but is not able to explain them; as for the widths, on the other hand, when asked to compare the two glasses he says *'There (C, at the top) it is little and there (B) it is big'*.

PAS (5; 6) shows progress over the preceding subject in that he anticipates an inversion in the succession of the levels and is thus intermediate between stages I and II: *'This bit* (first interval *of A*) *is way at the bottom here* (in B)', but he draws the intervals DHB the same size and does likewise for the DWB's. After he has verified the heights, level 4 is set for him and he is asked to anticipate the others. At that point Pas, contrary to Phi, takes into account the observations which he has just made and he draws intervals which decrease towards the top. But there is as yet no comprehension on his part since he continues to anticipate equal DWB intervals, and when we go to class C, he anticipates the height intervals DHC by simply reproducing the DHB's that he saw, without inversion. The WC's are again equal but when he observes the decrease of the WC's towards the top, he explains it by saying: *'Here the glass is small'* (= less and less wide).

There are many more examples of such reactions which are instructive in terms of the formation of functions. The point of departure is the assimilation to a scheme of action, in this particular case, the draining of a liquid in successive stages with constant intervals. The subject expects to rediscover this structure (given in *A*) in glasses *B* and *C* without taking into account their shapes nor their dimensions. Thus, as previously stated, the child anticipates a complete isomorphism, in spite of what he observed at the time of initial decanting. This isomorphism even encompasses the direction of the levels; the subject does not concern himself with the fact that although the water poured from *A* to *B* does first pass by the upper part of glass *B* it is nonetheless filled from the bottom up.

This initial function of complete equivalence or correspondence can lead in three directions due to the primacy of the assimilation of actions whose structure is simply transferred onto new objects. Two of these lead to the composition of new functions while the third results in their mere discovery by observation without bringing about immediate compositions. These directions are: first, the search for causes together with the coordinations they presuppose which are really based on the object; second, the operatory composition based on the coordinations of the actions of the subject; and finally, the abstraction of the dependences which were not perceived until then and which maintain their status of functional laws before being integrated into causal or operatory systems.

At this stage I the search for a causal explanation is still very limited which is the reason for the instability of the newly observed functions. Nevertheless, it is not inconsequential and the intermediate case of Pas shows how a subject begins to progress from the time such questions are asked. He comprehends (or remarks, contrary to Phi) that the water is not retained at the entrance to jar *B* and necessarily flows towards the bottom, and he makes the levels and their intervals start from the base of *B*, inverting the direction of the spatial correspondences between the *HA*'s and the *HB*'s or *HC*'s.

Operatory compositions are likewise limited to this level of evolution, resulting among other things in the non-coordination of the dimensions *H* and *W*, etc. What remains therefore are empirical observations during the verifications of the intervals *H* and the widths *W* which impose the discovery of new functions: the decrease in the intervals *HB* as a function of the widening of jar *B* from the bottom to the top, the increase of widths *WB* linked with same factor, etc. The salient fact which this stage I permits us to establish is that such functional dependences can be observed and grasped as dependences without being understood causally and due precisely to this lack of comprehension and to the consequent lack of composition, they remain unstable and are either rapidly forgotten or maintained as they are but are applied to situations for which they are not suited. This is why Phi, although he has observed the unequal intervals *DHB*, does not take the *DHC*'s into account while Pas carries the same intervals over to the *HC*'s but in decreasing order from bottom to top, without inverting them into increasing intervals.

This non-comprehension and the limited coherence of the functions thus recently discovered by observation no doubt stem from the absence of causal composition. In fact, while it may seem simple to understand that the same amount of water poured from *A* to *B* or to *C* will result in decreasing heights and increasing widths when the jar flares out towards the top and that the inverse will occur when it narrows, this comprehension of the cause and effect relationships between the shape of the jar and the levels of water which it contains presupposes a certain number of prior conditions, such as the non-dilatation and the non-compression of the liquid, i.e. the conservation of its quantity and of the volume which it occupies. At this stage I, the subjects have not yet acquired these notions, for lack of operatory structures, and their causal interpretation is thus affected. When Phi declares, for example, that the intervals *DHB* narrow at the top of glass *B* because 'there is not enough space' or because we put 'too much water' in the bottom, he knows that the intervals *DHB* correspond to equal quantities in *HA*, although the same quantities may not be exactly equal in *HB* because the glass is different: that is why the child thinks he should have managed to obtain equal *DHB*'s as drawn by him, and when this is not the case, the only thing to do is to accept the fact without understanding it (as happens in the conservation experiments when the child observes that the levels change with the shape of the glass). Therein lies the discovery of new dependences: the heights *HB* vary as a function of the shape of the glass, but since they are not causally or operatorily understandable, these functions are non-composable, and for lack of being composed into a system, they remain fragmentary and inconsistent.

3. STAGE II

In the subjects of this level we will simultaneously observe the discovery of the functions at play in the apparatus and their more or less correlative progressive composition:

JAC (7; 3) observes that in *A* the *DHA* intervals are equal. 'What does the water at the bottom do? – *It rises.* – When we pour a little bit (one *DHA*) how will it be (in *B*)? – *It is higher.* – Why? – *Because there it is thicker* (diameter of *A*) *and there it is thinner* (thickness of *B* = 1 cm).' We can thus see that the evaluation of the *DHB*'s is from the outset a function of the coordination of two dimensions, although at first Jac forgets the third and draws the *DHB* intervals as being equal although they are higher than the *DHA*'s: '*Because it's all the same thing* (in height)'. But having anticipated *DHB* intervals which are too big, Jac cannot fit them all on her page and when she starts over again with smaller *DHB*'s she realizes that she forgot a dimension: the *DHA*'s are equal '*but not here (DHB), because it's always leaning* (= the sides of glass *B* are at an angle and are not vertical)'. She then draws *DHB* intervals (starting from the bottom of *B*) '*smaller ones then larger ones. — Why? — Because it becomes larger and larger* (= wide)'. In other words, having understood that the heights of the *DHB*'s depend on the widths, Jac first of all hypothesizes that there is a direct rather than an inverse relationship between the variations of the *DHB* intervals and the increasing widths. The anticipation of the latter is on the other hand correct since it is a question of a progressive increase from bottom to top, but the *DWB*'s are still not understood. We then pass on to the verifications: '*Oh, no!*' says Jac of the *DHB*'s '*they are less and less thick* (less and less high). — Why? — *Because there* (bottom of *B*) *the glass is less thick* (= *less wide*)'. Jac thus understands the inverse relationship between the widths *WB* and the intervals *DHB* so well that she can find the position of a *DHB* by '*looking at the width, as it is*'. As regards the explanation of the decrease of the *DHB*'s as a function of the increase of the *WB*', Jac does not provide a justification in terms of compensations, but says '*because there is less and less water*' which is no doubt only an abbreviated form of the best explanation that Jac will give regarding the *DHC*'s.

As regards the *DHC* intervals, Jac begins by predicting their gradual decrease from bottom to top as she did for the *DHB*'s, but the prediction of the *WC*'s is correct: '*It is higher and higher* (height of the *DHC* intervals *because the glass becomes smaller and smaller* (= less and less wide) ... *It's smaller, so it rises more ... because it becomes smaller* (= less wide), *then the syrup goes higher*'. She has therefore grasped the compensation of the heights and the widths in terms of the *DHC*'s and *WC*'s without always understanding the *DWC*'s.

MAR (8; 5) begins by ordering the *DHB* intervals from top to bottom although not in the correct direction, but when we ask her where level HB_5 is in relation to the corresponding level in *A*, she immediately corrects herself. However, she still carries over *DHB*'s which are equal to each other and to the *DHA*'s. Then, observing that the total heights do not coincide in *A* and in *B*, she draws *DHB*'s which are different from the *DHA*'s but are still equal to each other. The intervals separating the widths are also equal to each other and the widths *WB* are immediately conceived as a function of the heights: '*When I made the height, then I saw the width*'. Upon verification, Mar right away understands the relation: '*When it is wider, the syrup rises just a little bit*'. For jar *C*, Mar right away declares: '*With that, it's just the opposite. At the top, it's smaller, at the bottom it's wider, so it will rise less — at the bottom less, at the top more*'. Her anticipations for the *DHC*'s and *DWC*'s are correct and when the widths are hidden so that only the heights are visible, Mar can determine which jar is involved: '*At the top it slows down a little (for B) and here (C), it rises more quickly*'.

STE (9; 5) first of all predicts the equality of the *DHB*'s '*for the same reason as there (DHA's)*' but in marking his first levels he admits '*it's difficult*'. There is also equality between the *DWB*'s. Upon verification of the *DHB*'s he immediately realizes his mistake: '*Oh! I get it, because here it is thinner and there it is wider and the water is thinner and there it is wider and the water has more space to spread out in: that depends on the amount of water*'. He then concludes that the *DWB*'s will also be '*increasingly smaller*'. For the *DHC*'s '*it will also be opposite*' with the intervals increasing. But for the *DWC*'s Ste predicts a decrease, confusing the widths *WC* and their differences *DWC*; upon verification, he exclaims: '*Oh! That's the shape of the vase, there is less space thus the water must rise*'.

EDA (10; 8) in spite of his age also foresees that the *DHB*'s and the *DWB*'s will always be the same since there (*A*) it is always the same. Based on the verification, he makes an exact anticipation for *DHC* and *DWC* because '*that's the opposite*'.

MIC (11; 3) reacts like Eda and after verifications and good anticipations for *C*, answers the question: 'When are the differences equal? – *When* (the container) *is square*'.

As regards the anticipations of the differences in the heights *DHB* or widths *DWB* in jar *B*, little progress has been made over stage I. On the other hand, the comprehension of the functions in play after observations and verifications marks a clear progress whose different aspects are instructive in terms of the elaboration of the structures of functions.

At the level of anticipations, the only real progress is the generally immediate comprehension of the fact that while the levels of the liquid *HA* follow each other from top to bottom in jar *A*, they succeed each other from bottom to top in the containers *B* and *C* since these are being filled. On the other hand the anticipation of the *DHB*'s or *DHC*'s and *DWB*'s or *DWC*'s always reveals the primacy of the strict isomorphism, these differences being considered equal among themselves in *B* and in *C* as they are in *A*. There is however in certain cases the beginning of a differentiation, as with Jac (7; 3) who indicates higher intervals in *B* than in *A*, because while *A* is cylindrical, *B* is only 1 cm wide, but these larger intervals remain equal.

In order to pass from these false anticipations to exact functions, the first possible path is that of causality, which already takes into account the fact that the child attributes to the filling of *B* and *C* the opposite direction from that of the draining of *A*. In fact, since causality constitutes a system of transformations and of concomitant conservations, it leads, at this stage, to new constructions elicited by the question regarding why the *DHB*'s vary, as well as by the experimental observations made upon verification. Let us recall that in stage I this question could not be answered due to the absence of the notion of the conservation of the quantity of liquids. This notion of conservation is acquired by children between 7 and 10 years of age, due in particular to the compensations discovered between the quantitative variations in the shape of the liquid for the two or three dimensions involved. Therein lies the process which we find here but with the emphasis placed on the variations and the reasons for them, in other words, on the functions and on the causes of the dependences which they express. In fact, the major progress of the subjects of this stage, once the observations have been made, is that of looking for the cause of the variations observed, by invoking almost immediately the interactions between the shape of the jar and the quantity of the liquid insofar as quantity is conserved.

This conservation is not explicitly affirmed, assuredly because first of all we do not pose this problem, and above all because, once acquired, it is taken for granted and does not require deliberate formulation to be used. But it happens that the child refers to it: 'It depends on the quantity of water', says Ste for example, thereby understanding that the same quantity which passes from an interval *DHA* to an interval *DHB* can thus 'be spread out' because it has 'more space' in width. But this same reasoning occurs implicitly in all of the other subjects who, accepting the conservation of quantity without question, then appeal to the causal dependence uniting the shape taken by the liquid to that of the solid container which imposes it in order to find the reasons for the variations of *DHB* and of *DHC*. As all of these subjects say, each in his own way, these intervals increase in height when the jar shrinks in width ('then the syrup goes higher', Jac) and decreases in height when the jar widens.

Of course this does not mean that causality as such is the source of functions. It does nevertheless seem that in the cases in which the heights and widths of the liquids have been thus observed and understood as functions of the shape of the jar, this functional dependence takes on in the eyes of the subject a physical significance whereupon these covariations which are found within the object itself are susceptible, together with the conservation of quantities, to being integrated into a causal system. (The system becoming causal only in so far as it is a system.) The system thus constituted acts in turn on the functions which constituted it, consolidates them, and above all confers on the dependences which they express the sense of 'connections' and not only of 'conjunctions' (to use Hume's vocabulary), i.e., an explanatory significance and no longer only a lawful one.

But the causal dimension of the functions elaborated at this stage is far from being the only one and it is necessary to correlatively focus on their operatory dimension which manifests itself in several ways: first in the actual constitution of the notion of a spatial interval as a straight line segment between two end-points (which is intuitively immediate in the case of the *DH*'s but less so in that of the *DW*'s); second, in the serial ordering of the increasing or decreasing intervals and above all in the direct or inverse serial correspondences; finally, in the composition of the relations which makes possible the attainment of the form of reciprocity comprising the compensation between the broadening of the width and the shrinking of the height, or vice versa. This last operatory composition intervenes in the construction of the notion of conservation and consequently in the system of transformations and integrated conservations which we would consider characteristic of causality. It suffices to say that there is an interdependence between the causal dimension and the operatory dimension of the development of functions at this stage, but that goes without saying if causality consists of operations attributed to the object on the model of the operations which characterize the actions of the subject.

Finally if the functions belonging to this stage are thus constituted according to two correlative dimensions — causal and operatory — there still remain functions imposed by observations made at the time of the experimental control which are neither causally 'comprehended' nor operatorily 'constructed'. This is the case with the difference in the DWC widths for jar C where the widths themselves are decreasing (from bottom to top) while their differences are increasing. As long as the differences of the intervals are envisaged in spatial terms and can be immediately intuited, as in the case of the easily orderable DH's, there is no problem. A problem does arise with the emergence of the abstract idea of variations, namely as it relates to the DWC's where the intervals are to be subtracted from the total original width instead of being added as with the DWB's. In this case, the child is naturally forced to accept the function but only by virtue of factual or lawful observations, and where these are concerned, we cannot speak of anticipations nor of real comprehension.

4. STAGE III AND CONCLUSIONS

In the third stage the subjects finally go beyond post-observational interpretation and successfully anticipate the DH's and the DW's and they also comprehend them as variations of variations or as functions of functions and no longer only as a geometric intuition. But it goes without saying that this progress does not take place suddenly and that there are many intermediary positions between stages II and III:

GIL (10; 0) exemplifies one of the intermediary positions. He immediately anticipates the decrease from bottom to top of the DHB intervals: '*At the bottom there's a bigger one (top) because the glass is less wide than at the top. After that it's smaller because it becomes wider and the syrup spreads along all of it*'. On the other hand he says that a ruler would be useful for measuring the DWB's on the glass and thinks that '*the little bits* (intervals) *are always larger because that becomes wider*' which is thus a confusion of the DWB's with the WB's, thus of the widths themselves with their differences. But after actual observation, he understands his error '*because this width there is larger than half of that one, thus it has to be a smaller bit which we add*'. As for glass C, he uses what he has just understood and anticipates the progressive growth of the DWC's and of the DHC's, inverting the functions of glass B.

LIP (10; 0) on the contrary completely attains stage III: he immediately anticipates the decrease of the *DWB*'s as well as the *DHB*'s, and tries to quantify these differences in terms of three-fourths, halves, etc. For glass *C* '*It is exactly the opposite of the other*' and the *DWC*'s like the *DHC*'s are '*larger and larger*' because that becomes '*each time less wide*. – Aren't they the same? – *No, larger*'. Even when the widths are hidden, Lip says that we can judge which is jar *B* or *C*, which he does in terms of speeds: '*I think we can still see it. Here* (*C*) *it will take longer to go up and afterwards it will go faster. It's the opposite there* (*B*) *but they will both be filled at the same time*'.

These cases, which occur primarily from age 11–12 on (level of the propositional operations) are remarkable in their use of operatory deduction and their attempts at metric quantification. But the physical or causal aspects of the functions in play do not for all that disappear. Lip gives us a good example of this when he expresses the intervals in terms of speeds and introduces the notion of the conservation of time which is natural in reference to the draining of the liquid from jar *A* taken as a whole (as opposed to successive intervals) but which presupposes an interplay of compensations between the initial and final speeds of the filling of jars *B* and *C*.

Generally, in comparing the successive stages I to III of this experiment, we observe a clear evolution of functions starting from the constitutive functions involved in stage I to the constituted functions of stage III.

Constitutive functions already exhibit the characteristics of functions although they are in fact manifested prior to the first concrete operations which emerge at age 7–8, thus before operatory reversibility and conservations. We could consequently compare them to simple relations because there do exist preoperatory relations or prerelations (without coherent compositions), but they are essentially different in that a relation can only result from a comparison. '*A* is larger than *B*' or '*A* is equal to *B*' are relations because they exist only from the moment a subject compares them, while in and of themselves *A* and *B* can only be equal or unequal, i.e. their sizes, which no doubt exist objectively, lend themselves well to comparison, but only in the same manner as a mushroom lends itself to be eaten and digested or on the contrary resists due to its toxicity; however, and therein lies the crux of the matter, *A* and *B* cannot compare each other in and of themselves. On the other hand, the 'dependence' expressed by a given function such that the level of the water in *B* depends on the quantity poured from *A*, is real and objective: though it results from the action of pouring *A* into *B*, this action, which in this context is executed by a subject, could easily be effected without him, thus between natural objects and not due to human intervention. Thus the source of a constitutive function is an action, no doubt at the outset an action of the subject but one which is an action of one object on another in its material as well as conscious aspects.

Linked as they are to actions whose schemes of assimilation they express, functions not only precede operations, but even encompass them, since operations derive from the actions of the subject or, more precisely, from their general coordinations (which ultimately results in closer and closer connections between certain functions and operations) and since not all individual actions are transformed into operations. This explains the general and elementary character of constitutive functions which express the dependences between the result of an action and the action itself, in a form which is generalizable (in the scheme of assimilation) and not only perceptual and real.

The simplest constitutive functions thus express a dependence centered on the possible equivalences between the action and its result. This is in fact what we observe in stage I of this experiment as well as in the initial anticipations of stage II when the child expects that the stages of the action of emptying jar A will be the same as those of the action of filling jars B and C including (in stage I) the direction of the successive levels. This is the fundamental characteristic of constitutive functions of correspondence, but we should also note that these correspondences, being pre-operatory, are essentially qualitative. Early experiments on the conservation of numerical sets showed us that there exists a visual although non-operatory correspondence such that two rows of tokens are considered equal when the elements on one are placed directly across from those of the other, but this equivalence disappears as soon as we eliminate the points of contact. This visual correspondence thus constitutes a function but does not yet constitute an operatory structure. It is the same characteristic which we find here in stage I. The anticipated correspondence remains qualitative and even figural in nature, but as soon as the observations contradict the anticipated situation, the subjects disavow the conservations and understand neither the compensations nor the new correspondences imposed by the facts. On the other hand from these initial qualitative functions to the full comprehension of stage III, we witness a progressive quantification; first by simple compensations which are logically deduced between the heights and the widths and then by attempts at metric quantification which appear spontaneously in stage III.

A second lesson to be learned from our results is that the formation and composition of functions adequate to the situations presented cannot be separated, which is of course natural to the extent that qualitative elements are inserted into the quantitative structures which complete them, since while the qualitative is immediately perceptible, even non-metric quantifications presuppose a construction. These initial compositions, whose progress is visible through stage II, seem to display a double nature: on the one hand there is a causal composition which is naturally imposed in that functions express the schematism of actions and since these actions are causal in nature; on the other hand there is an operatory composition, which permits the coordination of functions f and their inverses f^{-1}, thereby resulting in the attainment of compensations, etc. Only those functions which are observed but are not yet 'comprehended' nor assimilated (to be subsequently composed such as the DWC's) escape this double causal and operatory composition, thus remaining a collection of simple general facts or 'laws' instead of being converted into structured wholes.

But once the operatory structuring of the functions has been initiated (in its dual aspect of logico-mathematical structures or of the subject on the one hand and of causal structures or operations attributed to the object on the other) it sooner or later leads to the multiplication of structures. From the logico-mathematical point of view, the clearest example of this offered to us by the above results is that of the passage from functions expressing simple covariations (width and height) to derived functions bearing on the variation of variations, in this particular case, on the differences between the intervals superimposed on the intervals themselves which up until that point were the only ones comprehended. The gradual transformation in the present situation of the constitutive functions of stage I to the constituted functions of stage II and above all III is precisely this type of superimposition of structurings.

How then can we explain this discovery of the functions of functions and how can we conceive the relations between the physical aspect of functions and their aspect of operatory construction? There can be no doubt that variations exist objectively and that covariations also reveal dependences given by the facts independently of the subject. Thus there is no reason for this not to apply to variations of variations and it is no coincidence that at the beginning of the infinitesimal calculus, differentials and derivatives were conceived as physical quantities modified in time or as the limits approached by these quantities through a temporal process. Must we then admit that functions and functions of functions are drawn from the objects themselves by simple abstraction and that the elimination of time in the history of mathematical notions (see in this respect Grize's interesting thesis)[3] is reduced to the passage from the physical object to the schematized object in Gonseth's sense? In a sense, such a process certainly intervenes. However the initial difficulties encountered in the interpretation of the infinitesimal calculus due to this realism and to ignorance of the fact that the infinite is not a physical but an operatory notion demonstrate sufficiently that something more intervenes: even when the subject starts out with a physical given in spatial or functional domains, he comes to reconstruct it and these reconstructions or constructions are necessarily and sufficiently explained by the coordinations of his actions or operations. In the preceding reactions of stage I to III we observe the beginning of such an elaboration. After having considered the level *HB* or the borders of the widths *WB* only as ordinal elements (stage I), the subject constructs the notion of the interval *DHB* or *DWB* as an element comprised between borders (stage II): he can then abstract, by means of a more or less hyperordinal comparison, laws of increase or decrease which introduce the differences between those intervals which have become measurable (stage III). The variations of variations are thus isolated but by a reflective and no longer simple abstraction, from the same process of coordination which makes possible the completion of order by quantity, the later being a true construction in contrast to the solely perceptual qualitative givens. (Order itself presupposes the action of being placed in a functional relation but is closer to the givens than are the various subsequent constructions on the scale of logico-mathematical operations where each one takes as the object of its progressive constructions the results of the preceding construction.)

To sum up, the notion of the variation of variations does in fact correspond to a physics but in addition presupposes an algebra (and sooner or later an infinite algebra) which goes beyond and enriches the former with structures which are not given, but which are to be constructed.

NOTES

[1] With the collaboration of E. Schmid-Kitzikis.
[2] With regard to these unidimensional estimates, see note 7 in §2 of Chapter 3.
[3] As used by Papert in *Études*, Vol. XX, pp. 15–17.
[4] Grize, J. B., *Essai sur le rôle du temps en analyse mathématique classique*, Neuchâtel, 1954, 106 p.

PART II

THE QUANTIFICATION OF CONSTITUTED FUNCTIONS[1]

The five experiments presented in this part (chapters 8 to 13) were not conducted simultaneously but rather consecutively over a three-year period throughout which time we were able to refine our approach to the study of functions thanks to discussions in meetings and symposia held at the Center. Nevertheless, the five experiments presented here have a certain unity of outlook, i.e. they place the construction of proportional relationships within the context of the search for the law of functional variation.

1. DESCRIPTION

We were able to refine our experimental techniques by formulating the following questions:

1. Is the idea of function already contained in the attempt to establish a correspondence among the relevant variables? If this is so, how does the child choose, from among several given characteristics, those which are functionally related? Is it enough, in order to isolate the law of variation, to simply construct sequences of sizes and link the corresponding states in a one-to-one correspondence?

2. To what degree does the establishment of a relation between the sizes and their variations facilitate the inference of a causal relation of cause and effect?

3. Is the discovery of the functional law of the same nature as experimental induction? Is the process of experimental induction, i.e. the passage from the observation of facts to their generalization, in keeping with the discovery of the continuity of covariations in the function?

4. Does the comprehension of a functional law imply the deduction of its application, thus the need to introduce a metric? Starting from covariations, how can the child arrive at proportional relationships? Is it in the attempt to set up proportional relationships that the child deduces the laws of covariation?

2. EXPERIMENTAL HYPOTHESES

Before presenting each of the following experiments,[2] with its specific problems, technique and results, we would like to indicate very briefly how they are connected.

In *Experiment I* (Chapter 8), the function is contained in the establishment of a relation of compensation between two sizes of the same type (two segments, the height and the width of a rectangle). Establishing a relation between these two elements may be difficult due to the reciprocity of the compensations, i.e. opposite variations, when carried to their extreme, result in the disappearance of one of the segments.

In *Experiment II* (Chapter 9), perception is made to intervene so that we may study its role in the discovery of serial regularities and determine the degree to which the perceived givens do or do not facilitate the establishment of a relation between the two methods of ordering used.

In *Experiment III* (Chapter 10) causal consequence is introduced by making a circle (the wheel of a car) revolve. The distance covered by a revolution of the wheel is a function of the size of the wheel. The problem is to find out how the child determines the size from among the linked characteristics: radius, diameter and surface area.

In *Experiment IV* (Chapter 11) the child is asked to coordinate three variables: the size of the wheel, the distance travelled and the rotational speed. We suggest compositions of functions to the child. The distance travelled is both a function of the size of the wheel and of its rotational speed.

In *Experiment V* (Chapter 12) we originally wanted to study the diverse combinations of functions in a system of pulleys. The simplification of the apparatus led us to suggest to the children that they look for the equilibration of different weights through the inverse compensation of distances. Does the discovery of the constant K which is equal to weight W times the distance L (the length of the arm of the lever) presuppose the discovery of a functional law which links two magnitudes which are very different in nature? (That is to say, the weight can be effectively increased not only by adding weights but also by varying the length of the lever.)

NOTES

[1] By Vinh Bang.
[2] We are deeply grateful to N. Abramovitch, C. Castro, U. Steeb, V. Zaslawsky, R. Bréchet, A. Körffy, I. Monighetti, and S. Uzan for their collaboration in this research.

CHAPTER 8

THE FUNCTIONAL RELATION BETWEEN THE INCREASE AND THE DECREASE OF BOTH SIDES OF A RECTANGLE HAVING A CONSTANT PERIMETER

The Transformations of the Perimeter of a Square

1. DESCRIPTION[1]

If we consider a function solely as a relation between two sizes where the variation of one results in the variation of the other in the same proportion, then the study of the representation of the double compensations involved in the transformations of a square with a constant perimeter into a rectangle would provide a means of exploring the problems posed. The lengthening of the base of the rectangle is the consequence of the decrease of its height. Can we speak of the comprehension of a functional law as soon as the child has grasped this implication? Or rather will the intervention of a metric be necessary to quantify the covariation between the two sizes, the notion of function being acquired only from this moment on? This would presuppose a search for the size of the variation which is subject to a predetermined law of progression because it is linked to the covariation of the other size.

APPARATUS AND TECHNIQUE

(A) *Apparatus.* – A green cotton thread, 40 cm long, is placed around 4 pins so as to form a square having sides 10 cm in length. By moving the 4 pins, we can transform the square into a rectangle. The progressive decrease of the height brings about a progressive increase of the width. Ultimately, we obtain a double thread held by 2 pins.

(B) *Technique.* – The square (A, B, C, D) formed by the thread is placed on a sheet of paper fixed onto a pegboard. The thread follows the contour of a previously drawn square to be used as a frame of reference. The sides AC and CD coincide with two 35 cm long lines which represent the x and y coordinate axes (Figure 11).
　　The subject is first asked for his predictions and is then asked about the transformations which have just been effected. Next the child compares his predictions with the

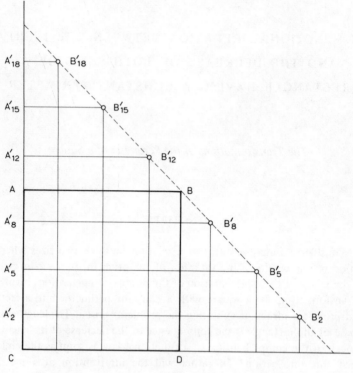

Fig. 11

actual results and explains why his prediction (previously marked in the pegboard by pins of different colors for *A, B, C, D*) should agree with the correct result. The questions include three parts:

1. The first phase relates to the decrease in the height. The experimenter successively moves *A* to *A'* (so that the ordinate goes from 10 cm to 8 cm, to 5 cm, to 2 cm and finally to 0 cm, where *A* and *A'* coincide). The child is free to predict the positions of *B'* and *D'*. The order adopted by the child in his prediction, i.e. first of all placing *B'* and then inferring *D'* or vice versa, could provide us with information regarding the steps involved in the deduction of a functional law.

2. When the positions successively taken by *B* are marked and corrected, we keep the (red) pins B'_8, B'_5, B'_2, B'_0, in place where they serve as points of reference. We ask the child to predict the transformations which relate to the decrease in width. Point *D* will successively take the positions *D'* (at 8, 5, 2 cm until *D'* is finally brought back to *C*). The transformation is thus symmetrical and the problem at hand is to find out if the child can profit from the experiment itself and from the observation of the results obtained in the first part of the test.

3. In the third part we request an explanation of the positions of *B*. Why are they located at an oblique angle of 45°? Would point *B* (if the pin were to pierce the thread,

i.e. remain fixed to the thread dividing the perimeter into two equal parts), always occupy the same place in the perimeter during the transformations of the quadrilateral? In other words, it would be interesting to find out whether the child reasons on the basis of the constancy of the entire perimeter or on half of it. This would be equivalent to operating on the displacement of a point which divides a line segment into two parts.

Note. – In the interest of clarity, we will call the pins which form the angles of the square A, B, C, D, C is also the origin O of the coordinate axes (x, y). The displacements of A are designated by A'_8, A'_5, A'_2, A'_0 and by A'_{12}, A'_{15}, A'_{18} and A'_{20}. The subscripts of the A''s and B''s indicate the values of the ordinate in centimeters. The subscripts of the D''s correspond to their values as measured along the abcissa.

2. RESULTS

We questioned 41 children aged 7 to 13. The criterion adopted to differentiate the types of behavior was defined as the child's capability to grasp the following relation of logical implication: the variation of one of the sizes (the height) *implies* the variation of the other (the width). In order for the increase in the width to be understood as the *consequence* of the decrease in the height (for the covariation to be determined) at least two states of successive transformations had to be linked. The covariation existed because we kept the perimeter of the rectangle constant.

We describe the different approaches which we observed on the basis of this criteria. Although we were not able to establish a hierarchy of the behaviors as they evolved with age (such an ordering presupposes an adequate and in-depth analysis), we will nevertheless present them in an evolutionary order.

TYPES OF BEHAVIOR

(a) *Failure to link two transformational states.* – This non-linkage appears in both phases of the experiment: the prediction and the observation of the transformation effected. In the first phase, the child cannot conceive that a decrease in the height will result in an increase in the width. For example, when we move A to point A'_8, the child moves B down to the same level (by sight) but B'_8 still remains on the side of the square. In other words the width of the figure remains constant. At first glance, it would seem that the child doesn't understand the task. But, upon closer examination, we find that this refusal to go beyond the boundaries of the model-figure during a transformation occurs repeatedly. In fact, when we effect the first transformation, even though the child has observed that point B'_8 is located well beyond the point he predicted, he will not take this into account for the next transformation. For A'_2, the child moves B'_2 to the same level, but always on the BD side of the square. Shortening one side does not necessarily bring about a compensation lengthening the other side. This becomes even clearer when the square is made into a rectangle whose width becomes smaller and smaller. For each transformation, the child maintains the height constant, i.e. equal to that of the preceding transformed figure. Even more striking is the fact that in the final transformation, the child still refers to the height of the preceding rectangle. For example, when we move D'_5 back to D'_0, the child immediately places the pin in the place occupied by A'_{15}. We describe this behavior as a non-linkage rather than as a non-compensation because the variation of one term does not in any way affect the other. The idea of compensation presupposes a covariation even if at the outset this covariation is contradictory, such as when the decrease of the height brings about a decrease in the width.

(b) *Compensation in the same direction.* – Judging strictly from the criterion of success and failure, it would seem that this behavior is more primitive than the one previously described. In our attempt to find out whether the child does or does not have the concept of a compensation of the variables involved, we found that there seemed to be a logical inference present in this type of behavior, even though it was incorrect. Confronted with behavior (*a*), we asked ourselves to what degree the border of a figure would distort the idea of covariation in this study.

In this behavior, however, the child already coordinates states of successive transformation: the decrease of one side brings about the decrease of the other. If the child remains unaffected by successive corrections – since after each prediction we proceed to verify the actual result, the points of reference remaining in place – it is due to the primacy of the inference 'smaller implies smaller'. This is the functional consequence of the transformation. Thus a decrease in the height or in the width brings about a decrease in the other dimension. We also note the child's concern with maintaining the ordering of the sizes.

(c) *Compensation of the sizes from one transformation to the other.* – This compensation generally bears on two contiguous states of a transformation. The first condition is that the conservation of the perimeter must be taken into consideration 'because it's the same thread' and 'since the perimeter is always the same, only the width and the length vary' (Phy, 12; 0). Based on this the child establishes a link between both sides of the quadrilateral. The amount lost in one dimension is gained by the other. Can we speak of a functional relation when Edw (age 10), based on *intensive* compensation, argues: 'The width is smaller, thus it lengthens', or Nor (8; 10) who says: 'I measure the difference in the distance between the (initial) side of the square and the given point; this difference is added to the length of the square'? We feel that a functional relation must contain the idea of continuous variation concretely expressed in an extensive if not metric seriation, as is the case in our experiment. However, in the first example cited, the compensation is accomplished step-by-step and the child is not at all surprised when in the last transformation the decrease of the height is no longer fully compensated by the increase in the width of the rectangle. In the second example, all the transformations are based on the initial square. Once we ask the subject to take as the point of origin not the initial square but rather any correctly transformed rectangle, the child hesitates and does not apply the scheme of compensation.

This is why we describe this behavior of compensation by saying that it proceeds step-by-step. The child has not yet discovered a system for regulating any transformation whatsoever.

(d) *The search for a term by term correspondence of the compensations between two series of variations.* – This behavior appears when we ask the child to find a system which regulates the compensations. First of all, we note that the child is no longer satisfied with 'more or less'. In other words, to each A', there corresponds a single B', thus the need to measure: 'We must measure to know the exact place'. The child still has to find out which size to measure. For quite some time the child will continue to look for this 'exact' compensation starting with a given situation, in order to find the solution for the next transformation requested rather than establishing a law of construction for all situations. Thus, he begins by searching for a procedure. For example: 'For it to be right, we must push the pins two widths (two times the distance AA') farther in relation to the length' (Cla, 11; 3). Two (or more) times because at the final transformation, the distance CD is twice that of a side of the initial square. Carrying over segments such as these can lead to absurdities. Thus, the child seeks to add to the width not what was removed but what remains of the height $(AA' - AC)$.

The passage from the first part to the second, where the same questions (symmetrical situations) were posed, helps the child to grasp a law of progression or diminution and discover the double seriation.

(e) *Discovery of the law of covariation and consideration of the limits of transformations.* – In proposing the use of a 'buckling square' for this experiment, we wanted to help the child discover the law of construction by manipulations and observations of the transformation of the square into a rectangle. But more specifically, the alignment of the red reference points (the positions of B' after successive transformations) at a 45° oblique angle suggested as we had hoped the discovery of the pairs of points A' and D' by the projection of B' onto the x and y axes.

In summary, we were able to point out two types of behavior leading to the discovery of the law of construction. Some subjects discover regularity in 'stair-like compensations'. They do this by seriating the heights and the widths and then simply by equalizing the compensation. 'It is because they are in steps' and 'what was taken away above (height) is added below (width)' (Phi, 13). The others start from the idea of compensations. For example: 'To find the exact point (B'), it is enough to measure the side of the initial square and to add to the length what we took away from the height' (Dan, 12; 6). The segments carried over are those found on the x and y axes. The oblique line is thus obtained 'because the perimeter is constant'.

NOTES

[1] The experimental approach of this study is a direct result of the research we conducted on 'spatial conservations'. In those results we found certain data which is of interest for the new direction being taken by research on the comprehension of the notion of function. When a child is presented with an L shaped string with a constant length (a pin O separates the two perpendicular segments AO and OB), the questions relative to the understanding of the reciprocal compensations of the lengths, i.e. the decrease of AO and the increase of OB, or vice versa, are solved by 7 year-olds. The same children, when asked about the transformations of a square (with a constant perimeter) into a rectangle, are no longer able to compensate the decrease of the height by an increase in the length. The sequence of transformations is represented by a series of rectangles, which are nested one within the other, which results in a shortening of both sides at the same time. Does the *décalage* of the levels at which solutions are reached result from a difference between the representation of the displacement of two segments and the coordination of two series of sizes? It is possible to first construct an ordered sequence of heights and to then coordinate each height with a corresponding width. Correct solutions are not attained until about age 11–12. Vinh Bang and Lunzer, Eric, 'Conservations spatiales', *Études d'épistémologie génétique* Vol. XIX, Presses Universitaires de France, 1965.

CHAPTER 9

SERIAL REGULARITIES AND PROPORTIONS

1. DESCRIPTION

The apparatus used in *Experiment I* did not allow the concrete seriation of sizes. The functional relation was that of a reciprocal compensation of two magnitudes of the same nature, i.e. the length of a segment. But if these magnitudes are materially distinct, and if their progression occurs in the same direction (for example, larger implies further), will the search for a functional relation be facilitated? In addition, when these series of sizes are presented in a regular manner, would perceptions contribute to the correct attainment of the results? If this were so, would the coordination between two seriations be sufficient to result in the isolation of a relationship of proportionality? In order to study this set of questions we present the child with a series of circles to be placed on a series of sticks. We define in advance the role of the size of the circles: it is as if the same circle grew as it was displaced on the stick. The place of a given circle is thus a function of its size. We will leave the child free to choose the criterion of size: the surface area, the diameter or the radius.

Apparatus and technique

Apparatus. – We have a series of 10 circles (from C_1 to C_{10}) whose diameters increase regularly from 1 to 10 cm, and a series of sticks S which are 100, 75, 50, 25 and 10 cm long. (S_{100} stands for the stick whose length is 100 cm.) Each circle has a hole in the center. We can attach the circle to the stick by putting a pin through its center or remove it while keeping the pin on the stick (Figure 12).

Technique. – We show the child the series of circles and explain their regular increase.

1. We place C_1 at the extreme left of S_{100} and with a gesture we suggest that if this circle grows regularly and becomes C_{10} it would be placed at the other end. We then ask the child to put C_7 exactly where it belongs on stick S_{100} and to explain how he proceeded. Then, we ask him to redo the same operation, with the other sticks presented one by one, in the order 50, 75, 25, 10.

2. Using the same sticks, we ask the child to find where C_3 belongs.

3. Once the places for C_3 and C_7 have been determined, we ask the child to put the sticks in order on the table indicating to him that this is another way of finding the exact place where the circles belong.

4. Once an order has been established, the child is invited to correct his errors and to explain his reasoning. Regardless of the result, we indicate the correct solutions for C_7

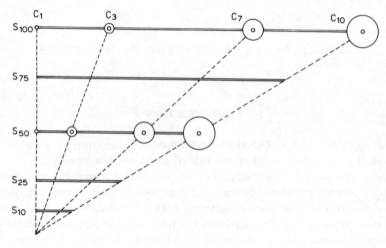

Fig. 12

and C_3 on S_{100} and on S_{25} and ask the child to correct the position of C_7 and C_3 on the other sticks.

5. From the explanations of the corrections, we pass on to the generalization of the procedure which is to establish a relationship of proportionality between the size of the circles and the length of the sticks. Another set of 'sticks' (strips of paper) in lengths of 140, 120, 80 and 60 cm are then introduced so that we may find out whether the child can apply the relationships of proportionality to these lengths.

2. RESULTS

The behaviors which we are going to describe are drawn from the reactions of 57 children aged 7 to 13. The most characteristic, those which can provide us with information regarding the comprehension of functions pertain to: (1) the relation to be established between the size of a circle C and its position on a given stick; (2) the establishment, during the seriation, of a relationship between the sizes of the sticks and the intervals between them in order to determine the relative position of the circles; and finally (3) the search for the proportional relation between the size of the circles and the length of the sticks.

Types of behavior

(a) *The search for the positions of the circles* on a stick S of a given length is illustrated by the following behaviors:

(1) *Failure to establish a relation*. Children age 7 (sometimes even age 8 or 9) tend to place the circles more or less at random. An order relation is attained when the child places a smaller one before a larger one. But this is only applied as a function of the series of circles and not as a function of the lengths of the sticks. Thus the child refuses to place the circles on the 10 cm stick because there is not enough space. For him the circles cannot be superimposed, they can at most touch each other (this in spite of the insistence of the experimenter).

(2) *The designation of an absolute position*. The circles are classified by size into three categories – large, medium and small – and in some cases, into certain intermediate categories whose existence already constitutes a degree of progress. Rob (8; 6) says: 'That's right in the middle, I'm sure', to place C_7 because for him 'there are big ones and little ones; we must leave space for them'. The same is true for Jos (8; 11) who reserves the central spot for C_7: 'because it's the "middle" and it must go in the middle' and for the same circle on another stick: 'In the middle, because it's medium'. At the beginning, older children give the same argument but subsequently provide more subtle explanations. Edw (10; 9) first of all places C_7 in the middle of S_{100}, then moves it towards C_{10} 'because it is a little smaller than C_{10}'. Dan (13; 2) does something quite different. He begins by pointing to C_7: 'The size of the circle is in-between the big C_{10} and the small C_1', but then he resolves all of the other questions by means of proportional relationships.

(3) *The search for a simple relation*. Towards age 9–10, we find either the search for qualitative compensations 'towards the circle closest to the circles of the same size' (C_7 towards C_{10}) for Mir (10; 8); 'it's closer to the big one (C_{10}) than to the smallest (C_1)' for Cha (11; 10), and 'I place it a little closer to the big circle' for Jos (11; 0); or the search for a seriation without the spontaneous coordination of the two sizes, circles and lengths. Mir (12; 10) asks to seriate the circles and adjusts them after having positioned them: 'All the circles must be put in place, otherwise it's too little or too big (this applies to the distances)'.

(4) *Discovery of a proportional relation*. Starting from age 12, the child is first of all interested in the size of the sticks and considers the relative place of each circle on each stick. 'The circle (C_7) must be placed more or less 3/4 of the way along each stick' declares Jac (13; 0) immediately guessing the size of the circle involved. Phi (13; 0) designates the approximate place of C_7: 'I put it here because the circles become larger and larger'; but he immediately considers the need for measurement 'we should measure, then divide to see how many (intervals) circles we can put'; he subsequently solves the question

by using fractions and not by proportions. The size of the circle is often determined by its rank in the series. 'I put the circle according to the place it occupies in the series and according to its rank in size' says Car (13; 5). Some children consider the regular seriation of the circles, such as when Chr (12; 7) says 'the circles must make a line (but when he demonstrates this he makes two lines, which are tangent to all the circles). Very few are able to establish a proportional relation as Geo (13; 0) does when he explains the principle: 'The ¹relationship between the parts of the stick remains the same because it depends on the diameter of the circles' and because the differences between the diameters are constant.

(b) *The seriation behaviors in this experiment are also informative in terms of the ability to isolate a functional relation.* – In effect, the intervals to be maintained between the sticks depend on the length differences between the sticks. Equal differences imply regular intervals. The S_{10} interval serves as a control. Prior to age 10, the seriation of the sticks does not in any way help the child solve the task of placing the circles. Except for two children age 8, none of the others see any relationships between the seriation of the sticks and the relative place of a given circle. Starting from age 10, the sticks S_{100}, S_{75}, S_{50} and S_{25} are placed in the correct order but the child does not change the interval between S_{25} and S_{10}. In spite of this, he discovers the correct way to place C_3. It is only starting from age 12 that the difference in the intervals between S_{10} and S_{25} and between S_{25} and S_{50} is noticed. 'Not always the same difference for this one (S_{10}), it is smaller', Chris (13; 5); 'a little closer for this one', Phi (13; 0). It is also starting from this age that the child draws the conclusion that circles of the same size always remain 'in the same (relative) place' on the series of sticks, and as we saw, either because it is the same fraction of 3/10 or 7/10 of the length of the stick, or because the distance is proportional to the diameter of the circles involved.

(c) *The establishment of a proportional relation.* – We have noticed that thanks to the seriation of the sticks starting from age 10, half of the children discover the correct position of the circles on each stick. What is striking is that these same children are not able to generalize this procedure for any stick of another length introduced later. This is true for the sticks which are 140, 120, 80 and 60 cm in length. This fact is consistent with the behavior of seriation pointed out previously. To find the sizes of these intervals as a function of the lengths of the sticks is to already apply a proportional relation. It is only starting from age 12 that children are able to successfully make this generalization for any stick length in order to find the exact place to be occupied by a given circle. But it is no doubt attributable to the apparatus that the child does not make this proportional relationship explicit as we might have wished. The children solve the problems by fractions, by the rank of the circle in the series and only very rarely by relating the covariation of the two sizes.

CHAPTER 10

THE RELATION BETWEEN THE SIZE OF A WHEEL
AND THE DISTANCE TRAVELLED

1. DESCRIPTION

The object of this task is the discovery that the distance travelled by a wheel (its circumference) is a function of the size of the wheel. Both sizes vary in the same direction, i.e. the larger the wheel, the greater the distance travelled, and vice versa. As a concrete example, we roll a wheel, the distance travelled being the result of this rotation. While the child can effectively observe the dimension of 'distance', how is he able to isolate the other dimension – the size of the wheel which can be calculated from the surface area, the radius or the diameter – to be coordinated in the system of variations? Finally, if the relation of the variations is constant, how does the child pass from simple correspondences to proportionality?

Apparatus and technique

We present the child with a car having 4 wheels or disks. The yellow front wheels measure 3.8 cm in diameter and 12 cm in circumference which is half the size of the blue rear wheels. The surfaces of the right front and rear left wheels have a stamp which leaves an impression mark once every complete turn. The car is rolled on a strip of paper 10 cm wide and 2 m long. We arrange for the blue and yellow marks to be side by side at the beginning. Thus the yellow marks coincide with the blue marks at the end of each cycle of turns (Figure 13).

The experiment is composed of 4 phases:

First phase: prediction of sequences of equal distances. – After the subject has observed the marks left on the strip of paper, the car is brought back to its point of departure. The child is asked to designate the position of the next marks (with tokens) as if the car was continuing its course up to the end of the road. The questions deal with the establishment of the correspondence between the distances and a given wheel and with the constancy of the distances between the marks of the same color (independently of the rotational speed).

Second phase: establishment of the relation between wheel and distance and quantification of the size of the wheels. – Here we have a series of wheels of different colors and colored paper strips corresponding to lengths equal to the circumference of the wheels. An identical set of disks and strips is made of white cardboard. We designate for our task: W_1 (diameter = 1.9 cm) and L_1 (circumference = 6 cm); W_2 (3.8 cm) and L_2 (12 cm), yellow wheels; W_3 (5.7 cm) and L_3 (18 cm); W_4 (7.6 cm) and L_4 (24 cm), blue wheels; W_5 (9.5 cm) and L_5 (30 cm); W_6 (11.4 cm) and L_6 (36 cm); W_7 (13.3 cm) and

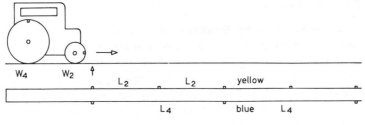

Fig. 13

L_7 (42 cm) lastly W_8 (15.2 cm) and L_8 (48 cm), the radii being drawn on one side of the wheels.

We place the yellow wheel (W_2), the blue wheel (W_4) and the small yellow strips (L_2) and blue ones (L_4) on the table. We ask the child 'by how much is the blue wheel larger than the yellow one?' in order to find out if he compares L_2 to L_4 to deduce the relationship between the size of the wheels or if he superimposes W_2 and W_4 to compare their surface areas. The questions center on the establishment of a relationship between the distances and the wheels.

Third phase: (a) *Prediction of wheel size starting from a given distance.* – We use the small colored strips and the white wheels. The prediction is made from $W_2 - L_2$ and $W_4 - L_4$ and from a given length L. The object of the requests for an explanation is a differentiation of the intensive deduction (a larger or smaller wheel than W_2 or W_4) from the extensive passage (by how much) to measurement (when one size is not an integral multiple of the other) or to proportionality.

(b) *Prediction of the distance from a given wheel.* – We use colored disks and small strips of white cardboard. Given $W_2 - L_2$ and $W_4 - L_4$, we propose a W for which the child must deduce the corresponding distance L. We ask him to justify his choice and the way in which he determined it according to an intensive criterion, by an extensive comparison, or by the consideration of the relation or the proportional sizes.

Fourth phase: functional law and proportional law. – This last part concerns the generalization of the law of progression. In order to facilitate the proportional construction of the two series of sizes, we presented a few subjects with the wheels whose radii had been marked, hypothesizing that they would thereby more easily isolate the relevant size, the radius, from among the other characteristics, the diameter or the surface area of the circle.

2. RESULTS

We questioned 116 6 to 13 year-olds. Due to the rather high number of subjects questioned in this study, we can analyze not only the typical behaviors but also characterize the stages of evolution. The experiment is composed of four phases. Thus, if the child seeks to relate the sizes of the wheels and the distances from the first phase, the questioning will focus on the search for proportional relations with phases 2 and 3 becoming instances of verification of the functional law.

1. Types of behaviors

(a) *Conservation of equal distances.* — Without this conservation, any subsequent comparison would be illusory. We consider that this notion has been acquired starting from age 8 when 66% of the answers are correct, even though the child does not always seek to use an adequate instrument of measurement (stick, thread, strip, etc.). By contrast, almost all of the children aged 6 (88%) foresee irregular distances, generally closer and closer together. While we have been successful (for example by repetition of the experiment) in helping most subjects discover this equalization of distances, there nonetheless remain 40% of the subjects at age 6 and about 20% at age 7–8 who continue the following behaviors which hinder them from grasping the solutions in the rest of the experiment. In the first one the distances are equal for the yellow wheel (W_2) and for the blue one (W_4). The distances may be of any size and only correspond to one turn of the wheel regardless of its size. In the next one, the behavior consists of the prediction of the comparison between two given distances instead of between one of the distances and the distances actually travelled by one turn of the wheel. The last one involves the rotational speed as if the distances were a function of the speed, i.e. the distances get smaller when the car goes faster. We will set this problem aside until Experiment IV. In the meantime, we tell the child that the car advances regularly and we make it move at a constant speed so as not to introduce one more variable!

Starting at age 8, the child understands that for each size wheel there corresponds a given distance which remains constant regardless of the car's movements.

(b) *Comparison between the sizes of the wheels and the lengths of the distances: different behaviors in the establishment of relations.* — When a correspondence is established between W_2 and L_2, on the one hand, and W_4 and L_4, on the other hand, the questions relative to the comprehension of the relationships which exist between W_2 and W_4 and between L_2 and L_4 set in motion different types of behaviors in the establishment of relations that can be classified hierarchically.

(1) *Intensive quantification.* — The comparison is based on a perceptual estimate without recourse to any measurement. The distance L_4 is judged to be larger than L_2 because W_4 is larger than W_2 by 50% of the children age 6 and 66% of those age 7. However, the functional relation remains intensive, which can be expressed as 'larger implies larger' and the child does not seek to

establish the relationships to find out by how much one size is larger than another.

(2) *Search for quantification.* – The first relationship established by the child aged 6–7 concerns distances. L_4 is judged on sight to be 2 times longer than L_2. The sizes of the wheels are estimated perceptually without there being any relationship between a wheel and the distance. W_4 is 3, 4, 5 times larger than W_2 or W_2 is 3, 4 times, etc. smaller than W_4.

When, starting at age 8 (96%) the child seeks to quantify the sizes of the wheels, we witness confusions which we can classify into three types as follows: (1) confusion in the comparison of the surface areas: the child displaces W_2 on W_4 or places W_2 in the middle of W_4 and concludes that W_2 is 3, 4, 5 or 6 times smaller than W_4; (the reciprocal of these relationships is not always respected: W_2 is 3 times smaller than W_4 but W_4 can be 4 times larger than W_2); (2) confusion of the parts and the whole: very often and even up to age 10–11, the child considers that W_2 is one time smaller than W_4 because he can fit W_2 into W_4 one more time (as L_2 on L_4), by superimposition; (3) confusion in the reciprocity of the relation wheel-distance: W_2 is judged to be smaller than W_4 by half but L_2 is estimated to be one time shorter than L_4 or W_4 one time larger than W_2 (superimposition) while L_4 is evaluated as two times larger than L_2 (without superimposition).

(3) *Search for a logical inference.* – Starting with the comparison of the lengths, the child seeks to 'justify' that W_4 is 2 times larger than W_2. This is why from ages 9–10, the relation wheel-distance is reduced to that of two sizes of the same nature: the length of the radius or of the diameter and the length of the distance.

It is for this reason that we draw the radii on the wheels and show them to the children who do not discover that the radius or the diameter determines the size of a circle.

We have emphasized our description of these behaviors because these orient the subsequent steps of the child, whether towards the search for proportional relations or towards the additive composition of sizes.

(c) *Prediction of the size of a term on the basis of the knowledge of the size of the other term by the establishment of a functional relation.*

(1) *Prediction of the size of W starting from a given L.* – We find the following typical behaviors:

(α) *Simple correspondence and failure to establish a relation among the givens.* The concept of function presupposes that the modification of one variable necessarily results in that of the other. In a simple correspondence, the child proposes to roll the wheel (or rather successively tries the wheels at random) to see which one corresponds to the given distance. Even more simply, he suggests making a wheel from the cardboard strip and then looking at the wheel of the same size. This response is most frequent at age 7 (40% of the subjects).

(β) *Establishment of a relation among ordered series.* – Judging by the sole criterion that a shorter distance implies a smaller wheel, 75% of the children aged 8 already give a 'qualitatively' correct solution. When we introduce a strip of length L placed between two others, one longer and one shorter, the child chooses a wheel W of intermediate size. On the other hand, when we give two intermediate choices, it is not until age 10 that we obtain the correct solution (50%).

(γ) *Construction of proportional relationships.* This cannot be achieved without recourse to measurement. Let us distinguish two behaviors which characterize the process of the application of a metric. In the first, the child seeks to add the differences of the distances to make these correspond to the same progresssion of differences of the wheels. The difference between L_1 and L_2 is the same as between L_2 and L_3, and the child seeks the same difference between W_1 and W_2 and then W_2 and W_3. If we give W_1, W_2 to find W_4 from the distances L_1, L_2 and L_4 the child reverts into the 'one time larger' type of reasoning and gives W_3 instead of W_4. The same failures can be found when we give two sizes of which one is not an integral multiple of the other, for example L_3 and L_4. 12% of the children aged 8 and 40% of those aged 9 exhibit this behavior which at first sight appears to be a search for proportionality. But the true level of the search for proportional sizes is found closer to age 13 (50%). At age 12, we find 20%. The child does not seek to compare the absolute differences but rather the relative differences even though he might say: 'Here it is the *same* and there also' meaning that the difference *between* the lengths L is the same as the difference *between* the wheels W.

(2) *Predictions of L from W.* – The order of questions (3a) and (3b) is deliberate. It is more difficult to make comparisons between the sizes of the wheels than to deduce the lengths of the distances. There is a clear *décalage* between ages 7 and 9. In other words, the same children regress in their behavior between the two phases of the experiment, a regression across all levels not related to any one type of behavior with the exception of precisely those 12 and 13 year-olds who are able to isolate the laws of proportionality.

Although the introduction of the wheels with the drawing of the radii facilitates the resolution of the problems, it also adds an element and reduces the construction to the simple coordination of two seriations of lengths.

2. Evolution of the behaviors

On analyzing all of the behaviors exhibited in this experiment, we can distinguish the following levels:

Level I (age 6–8):

The behaviors at this level are characterized by the perceptual evaluation of sizes, by the failure to establish correspondences between the wheel and its distance and between the two series of variables.

PHI (6; 11) foresees and marks the distances irregularly while affirming that they are equal and does not correct them when we ask him to do so. To justify the choice of the wheels, the child answers: *'I guessed'*.

PAT (6; 6) begins by drawing irregular distances then corrects them at our request and with our help. Prediction of W starting from L_1: *'The wheel which is one time smaller than the yellow wheel (W_2). –* How do you know? *– By looking at the wheels. –* The wheels or the distances? *– . . . I looked, one time smaller'*. In predicting W for L_2, he believes that the diameter is equal to the distance (W_6).

FAB (8; 7). 'How can we know that W_3 is three times larger than that one (W_1)? – (He moves W_1 8 times on W_3): *It is 8 times larger. –* Is that then the right wheel? *– Yes.*'

The child is often surprised at the lack of a common measurement between the two sizes to be related, but does accept the relation: 'smaller imples smaller'.

MART (7; 9). *'The yellow one (W_2) is two times smaller* (than W_4). – How do you know? *– Because this one (W_4) is larger than (W_2). –* And the distances? *– The yellow distance is 3 times smaller. –* Why 3 times? *– Because the yellow wheel is smaller. –* Could it also be 2 times smaller? *– Yes, 2 times. –* And the wheel? *– 3 times'*.

Exact quantification is lacking despite the appearance of a correct logical inference of the law of progression. *'If the distances are half, the wheels are also (Chr, 7; 11).* But for Anne (7; 10): *'The blue wheel (W_4) is 4 times larger* (than W_2) *and the blue distance is also 4 times larger than the yellow one'*.

Level II (age 8–11):

The progress evident at this level consists of the establishment of a relation among ordered series of sizes. Explanations are reduced to the level of empirical verifications and do not reach the logical deduction of the relations of the sizes. The introduction of measurement serves above all to justify the serial progression.

TRIK (9; 8). Prediction of L starting from W_8: *'Two and a half times larger than the distance D'.* To justify himself, he measures the diameters of W_4 and W_8, *'7.5 and 15 cm. That must be more or less two and a half times'*.

JOS (9; 10). Prediction of W starting from L_1: *'The wheel which is half of the yellow wheel (W_2)'.* She twice carries over the distance L_1 onto L_2 and chooses the wheel W_1. For justification: *'You must roll the wheel. –* How can you know that the chosen wheel is half of the yellow one? – (She doesn't know)'.

BER (10; 11) chooses for L_1, W_1. – 'Why this one? *– It's the smallest. –* Is there another way of verifying that it's the right wheel? *– By rolling the wheel.*' For the prediction of L from W_8, he measures the diameters of W_8 and W_4: '*7.5 and 15, it is 3 times larger than L_4.* – Are you sure? – *Yes*'. He chooses a distance which is too long. Justification: '*You must roll the wheel.* – And without rolling it? Try to carry it back 3 times the distance L_4 on the distance you chose. – *No, it won't work*'. He can no longer find the distance L_8.

We do not want to extend these examples of the search for corresponding sizes without bringing in the establishment of a proportional relationship in the sequences of values. The relationship between W_1 and L_1 is invariable for every pair of values: W_2 is to L_2 as W_3 is to L_3, etc.

Level III (from age 11 on):
It is from age 11–12 that we observe the use of the word relation (*rapport*).

MIC (13; 6). '*In relation to the blue wheel, it is three-fourths.* – How do you know? – *I compared it with the other distances.*' The solution W is deduced from the establishment of the relationship among the distances. For W_8: '*The new distance (L_8) is two times larger than the blue distance (L_4)*'. He measures the diameter of the wheels W_8 and W_4 to verify: '*This one (W_4) is half of the other (W_8)*'. He chooses distance L_8.
    The deduction of a desired size is made from the relations constructed from the series of the corresponding variables.
    MAR (13; 3). For L_8 spontaneously: '*It is 2 times larger than L_4, 4 times larger than L_2 and 8 times larger than L_1. I must look for a wheel which has a diameter double that of W_4.*'

Thus the function is understood and used as an application of a metric in proportional relations.

CHAPTER 11

THE ESTABLISHMENT OF A FUNCTIONAL RELATION AMONG SEVERAL VARIABLES: DISTANCE TRAVELLED, WHEEL SIZE AND ROTATIONAL FREQUENCY

1. DESCRIPTION

Our object has been to understand how the child constructs a functional relation from among several variables. In *Experiment III* we dealt with the coordination of two series of variables: wheel size and distances travelled. In the present experiment we will add a third, rotational frequency. We use the term frequency because the object of the task is the composition of the relative speed of two wheels during the same time period and not the estimation of their absolute speeds. Let W be the size of the wheel, D the distance travelled and F the rotational frequency; we thus have three linked variables, i.e. whether the yellow car A goes faster or farther than the green car B is a function of W, D and F. We will attempt to find out how the child establishes: (1) compositions of qualitative relations when one of the variables is kept constant, for example: $D_1 > D_2$ and $W_1 = W_2 \Rightarrow F_1 > F_2$ (D_1 is the distance travelled by the first car A, and D_2 that travelled by B); (2) reciprocal compensations such as when $D_1 > D_2$ and $W_1 > W_2$ and the subject is asked to find the relationships of rotational frequency between 1 and 2; and (3) the quantification of these relationships.

Apparatus and technique

Apparatus. – Two parallel grooves (track 1 and 2) are carved into a board which is 2 meters long and 10 cm wide. A yellow car (A) and a green one (B) slide in the grooves. On one end of the board there is an axis assembly which forms the axis for the pairs of pulley-like wheels whose diameters are 5, 7.5, 10, 12.5 and 15 cm. As the wheels are turned they reel in the string which thus pulls the cars toward the wheels. Yellow and green tokens are used to distinguish track 1 from track 2, and strips of colored paper are used to mark the distances travelled by each wheel (Figure 14).

Technique. – The questioning was clinical and followed this order:

What must you do
1. so that the cars will advance at the same time, together and parallel (side by side)? The object of this question is to find out if the child seeks to compose the variables from the start, or if he keeps them separate;

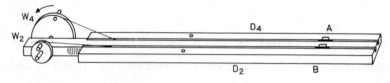

Fig. 14

2. for *A* to go faster than *B*?
3. (*a*) for *A* to travel twice as far as *B*? We ask this by indicating the marks instead of by explicitly stating the relationship of the lengths;
   (*b*) Same as in (3*a*) but here the question is aimed at the relationship of frequency. How many times faster?
   (*c*) for *A* and *B* to maintain the same rotational speed? This question necessitates the discovery of the relation between the wheel size and the distance travelled.
4. to find the relationships between the various wheel sizes?
5. for *A* to travel the distance *D* while *B* travels distance *D*/2 (without indicating the ratio)?
6. for W_1 (2 times) larger than W_2 to start and finish at the same time (remaining side by side)?
7. (*a*) for W_1 to travel twice as far as W_2 when $W_1 > W_2$ (ratio 1/2)?
   (*b*) The same as in (7*a*), but with the distances inverted such that $D_2 = 2D_1$.
   In some cases, we give *W*'s with other dimensions, or the ratios between the *D*'s are changed.

2. RESULTS

The results studied are drawn from 74 subjects aged 6 to 14. Our purpose was to distinguish the stages of the establishment of a relation among the variables involved. How does the child coordinate the relevant relations and above all how does he pass from the mastery of two variables to the command of the three variables linked in the experiment?

Types of behavior

(a) *Inability to coordinate two variables.* – We should note that in all of the questions, a bell sounds at the beginning and at the end so that the child may better understand the simultaneity and thus the identical duration of the displacements of *A* and *B*.

While all of the 6 year-olds answer the first question by saying that *A* and *B* advance side by side as long as the wheel sizes are equal, it is only at age 9 that they discover the second condition, i.e. that the rotational speed must remain identical. On the other hand, no child under age 12–13 proposed any other solutions such as that of varying the wheel size and the rotational speed.

But starting with the second question, we note interesting behaviors relative to the problem of the overtaking of one object by another. 6 year-olds insist that we must turn W_1 longer in spite of the fact that the bell has rung. 'I turn the yellow wheel all the way to the end but with the other, I stop before then' (Jac 6; 0). When one car overtakes the other the child is not able to grasp that they stopped simultaneously. This inability to coordinate the relevant variables is systematically found in 6 year-olds when the large wheel makes A advance more rapidly than the small one does B. Did (6; 1) makes both wheels turn at the same speed, but states: 'I turned the yellow one faster and the other one more slowly' because he perceives that A passed B. The same applies to attempts to solve the question when $W_1 > W_2$ and $D_1 > D_2$. 'We must', says Ber (5; 10), 'turn the yellow one faster' because $D_1 > D_2$. More precisely, Paule (6; 1) states: 'I turn the big one fast and the other slowly', because the large one has a greater distance to travel, but the large one still passes the small one. As we can see the overtaking is of prime importance and the child is not yet able to dissociate the two factors: wheel size and rotational speed.

As regards the distances and the sizes of the wheels, comparisons are limited to the relations 'greater than' and 'less than'. Even twice a certain distance is not grasped as such. All of the questions relating to quantification and to reciprocal solutions remain unsolved due to the inability to coordinate the variables in play.

(b) *Beginning of coordinations.* – The search for coordinations begins around age 8, although the coordination of variables does not arise spontaneously. It is necessary for us to propose either the change of wheels or the variation of frequencies.

If the child is able to establish relations between the relevant factors, it is because the effect produced by each one has been grasped. Thus the child must distinguish that a large wheel pulls the car faster, and also that with each turn of the wheel, the car travels farther than with a smaller one. With this dissociation of causes, the child is able to coordinate the variables involved little by little by qualitative compensations. When Mar (7; 9) says 'that the small one (wheel) must turn *more* to catch up with the large one' she does not yet know how many times faster. During these attempts at coordination we observe false compensations. When Did (8; 11) finds that he must make it turn 'one time faster with the large wheel', he has compounded two conditions where one would have been sufficient. To cover the same distance it is enough to make an identical wheel turn twice as fast, or to make the large wheel turn at the same speed. The same holds for Isa (8; 6) who, when she passes on to the compensations of the relations of distance, says: 'I make 4 turns with the large one and 8 turns with the small one (for $D_1 = 2D_2$)', which would be correct if we wanted $D_1 = D_2$. Instead of making an inverse compensation, she made a direct one. The same applies to Dom (8; 0) who finds that 'we must make 2 turns with the yellow one and a half turn with the green one'. One of the conditions suffices, both together are incorrect: either we need $2F$ for A and $1F$ for B; or we need $1F$ for A and $F/2$ for B.

(c) *Correct compensation for two variables, with the third remaining constant.* – Beginning at age 9, correct solutions are given for problems involving simple compensations. It is also from this moment on that the ratios 1 : 2 and 1 : 1/2 are quantified. We also note that the child spontaneously asks to vary the factors, for example to change the wheels to make them turn faster or slower. Thanks to the search for quantifications, the establishment of simple relations becomes possible; for example at age 7–8 we have observed instances of qualitative compensations: i.e. a *higher* frequency to advance *faster*. These differ from the following: 'When the little one turns twice the big one only turns once for the same distance' (Cat 9; 6) or 'We must make the wheel corresponding to the yellow turn twice as fast' (Fra 9; 10) for the question where $W_1 = W_2$ and $D_1 = 2D_2$.

(d) *Coordination of several relations at the same time.* – Between age 10 and 12, the child seeks to coordinate all of the variables. Successive answers contain fewer and fewer contradictions. For example:

GIL (10; 2). 'If I turn both wheels fast, but make them go the same number of turns? – *It's still the same, they will go the same distance.* – Question relating to the overtaking. – *I turn the yellow one faster.* He shows the exact distances predicted for A with $1F$ and B with $2F$. For $D_1 = 2D_2$ and $W_1 = W_2$: '*The yellow is double; I move the yellow one twice as fast because its distance is double that of the green one*'. For $D_1 = 2D_2$ and $W_1 = W_2$: '*The yellow is double; I move the yellow one twice as fast because its distance is double that of the green one*'. For $D_1 = 2D_2$ and $W_1 = 2W_2$: '*Together, at the same speed.* – W_1 is how many times larger? – *2 times*'. For the inverse question $D_1 = 1/2D_2$ and $W_1 = 2W_2$: '*At the same speed . . .*' The child begins to realize that he has committed an error of reasoning. – '*Oh! I must turn the large wheel two times faster than the small one*' and realizing his mistake again says: '*Half a turn with the large one and two with the small one*' which is correct. This error, however, is not committed by Dan (10; 0): '*I see, we need one turn with the large one and four turns with the small one*'.

This last question is the most difficult to solve because the large wheel has to travel a small distance. There is a double inverse compensation which is only solved towards age 12. The child thus coordinates the three relations, knowing that the car advances as a function of W, D and F at the same time. Here are some examples of correct solutions.

CRIS (12; 0): '*Oh, I take the large wheel which I place on the yellow car.* – The small one goes 2 times faster than the other because the large one is two times larger' (for equal distance) and for $D_2 = 2D_1$? – *Then the small one goes 4 times faster'.*

DANI (12; 0): '*We need twice as many turns with the small one.* – For $D_2 = 2D_1$? – *We need 4 turns with the small one and only one with the large one'.*

JAC (13; 4): '*Oh! I understand, we need to make half-turns. The medium one makes one turn and the small wheel one and a half turns'.* This last example shows that even when the wheels are changed the relations are reestablished by the child once a relation has been established among the variables.

CHAPTER 12

THE INVERSE PROPORTIONAL RELATIONSHIP BETWEEN WEIGHT W AND DISTANCE D (ARM OF A LEVER) IN THE EQUILIBRIUM OF A BALANCE[1]

1. DESCRIPTION

In the preceding experiments (II, III and IV), the functional relationship was directly proportional. This experiment comprises, in addition to the reciprocal compensation of the sizes involved in *Experiment I*, the establishment of an inverse proportionality. Our purpose was not to make the problem more difficult to solve but to understand how the child is able to establish a functional law by experimental induction. Given a balance arm in equilibrium with a weight W at a distance D, we can maintain the balance in equilibrium by modifying W or D. The problem consists in finding for a weight W hung at a distance D, the distances $D/2, D/3, D/4 \ldots D/n$ corresponding to $2W, 3W, 4W \ldots nW$, or vice versa. Although $1W, 2W$, etc. are discontinuous sizes, it is possible to establish the idea of continuous variation and also of the limit nW for a distance reduced to zero. Lastly, from the causal standpoint, the object is to find out how a child would know that a weight in equilibrium varies — without any modification in its intrinsic weight — as a function of the length of the arm of the lever. The functional relation is more natural but the comprehension of a physical causality becomes a pre-condition to the construction of the function.

Apparatus and technique

(A) *Apparatus.* – The balance is composed of a rectangular brass bar (4 mm x 20 mm) 48 cm in length, with a hole in the center (axis). Another bar with a base supports the beam by a pivot which passes through the center hole (Figure 15).
   The *weights* are 4 mm brass rods 4.5 cm long which can be linked end to end. The weight is thus translated in terms of length. A set of hooks is used to hang the weights from the beam.

(B) *Technique:* 1. *Presentation.* – Two pans A and B hung at an equal distance from the beam are in equilibrium. When we pour a certain amount of grain into pan A, the children observe that the bar tilts to one side. Almost all of the children tested pour an

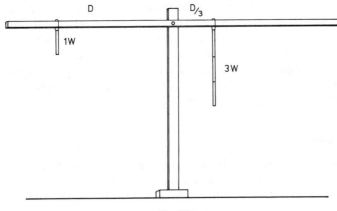

Fig. 15

equal amount into *B* to reestablish the equilibrium. Adding grain to *A* we ask the child to find another way to reestablish the equilibrium without adding grain to *B*. Since the child does not find the solution, we suggest that he move *A*. He is able to maintain the equilibrium by trial and error. By adding a certain amount of grain to *A* several times we make him discover that it is enough to bring the pan *A* progressively back to the axis to reestablish the equilibrium.

Sometimes, to better demonstrate the phenomenon, we ask the child to hold a yardstick horizontally by one end. We then place a book astride the ruler and make him observe that the book becomes heavier and heavier as we slide it away from the point of support, his hand.

2. *Questions.* – Two equal weights 1*A* and 1*B* are hung at about 20 cm from the axis and the child marks their position on the beam. 1*B* will remain fixed during the experiment. We add another weight to 1*A* (= 2*A*) and the child must anticipate the equilibrium position of 2*A*, verify it and give the pertinent explanations.

Once the position of 2*A* has been verified, marked on the pan (the marks at 1*A* and 2*A* remain visible), we pose the same questions for 3*A*, then 4*A*, 5*A*, etc.

We pass on to generalization, starting the experiment over with another position 1*B* (16 cm for example) and the child is invited to find the procedure for placing 2*A*, 3*A*, etc. precisely and without trial and error. As a control, we also make him do it in the inverse order, for example, from 2*A* to 1*A*.

The questions relating to the limit are then asked.

Note. – Using a parallel technique in which cardboard strips representing the arm of the lever are given, the child can each time cut out the lengths corresponding to the distances *D*/1, *D*/2, etc. He can compare both lengths directly: the distances *D* and the 'lengths' of the weights.

2. RESULTS

The overall results (for 65 children aged 6 to 14) were obtained using slightly different questioning techniques. The weights added are given either in the order 1, 2, 3, 4, etc. or in the progression 1, 2, 4, 8, etc. The distances are marked either directly on the beam or by making the child cut strips of paper which correspond to the distance D. On the one hand, we wished to know whether the child reasons more readily on absolute distances or on relative distances as he constructs the ordinal and then metric series of lengths; on the other hand, in certain cases we wished to avoid the application of an empirical scheme, e.g. doubling the weight corresponds to halving the distance.

Types of behavior

(a) *Failure to establish a relation among the variables.* — The child aged 6–7 is not able to grasp the interdependence of the sizes involved, above all the direction of their variation. To be able to successively equilibrate the scale in some given situations by trial and error is one thing, to isolate the relationship which exists between these empirically obtained results is another. The child does not therefore discover the direction of the variation, i.e. the progressive displacement towards the center of the beam, which results in a decrease in the distance when weights are added. The hesitation in the predictions, the trial and error, the displacement backwards or forwards of a weight without regard for the marks already made, show that the child at this age is not able to maintain an ordered relation and above all relations of connexity, i.e. the child acknowledges that $D/3$ and $D/4$ are smaller than $D/2$ but not that $D/4$ is smaller than $D/3$.

(b) *Discovery of a causal link and qualitative compensation.* — Qualitative compensations are not correctly expressed except when the direction of the decrease is maintained. The more weight we add, 'the more those (the distances) become *even* smaller'. In the arguments of Chan (6; 9), Bal (7; 1) as in those of Lor (7; 5): 'No (to reject the equality) because that ($D/5$) must be shorter than 4'. It should be noted that the increase in weights necessarily implies the decrease in distances. As the child displaces the weights each time a little more towards the axis, he quickly arrives at the limit. 'We can no longer add the weights (after 6 or 8) because we will pass to the other side (of the axis)', Mar (7; 1), Syl (7; 2), etc. Others say exactly the opposite. 'We cannot get there (to the axis)' say Fra (9; 2) and Pie (9; 7) who would like to maintain a certain distance for the maximum weight. These behaviors support the fact that the initial distance $D/1$ is not considered a base length, or a unit (because all of the other distances would vary as a function of the latter) from which the series of sizes of distances are constructed as a function of the weights. That is why more than 50% of the children aged 7 to 10 proceed with the ideas of compensation and of seriations of distances without arriving at a correct solution, since they are unable to represent the limiting bounds and to construct the sequence of sizes. The child seems to have grasped the causal relation: more weight implies a shorter distance. However, there still remain several stages prior to the discovery of the law of progression as it applies to this lever principle.

(c) *Search for a law of progression.* – 'The distance is smaller, I see it, but how much smaller, is really something else. . . .' This comment by Cla (11; 1) shows the true problem as posed by the child aged 10 to 12. The following examples are chosen to illustrate the stages of the construction of the law of progression.

(1) The children first of all seek to carry over the same interval of difference between $1W$ and $2W$ to place $3W$, thus placing it at the axis of the beam. When the child observes his mistake, he seeks to apply half of the distance $1W - 2W$. It should be noted that the idea of adding another fraction of distance is contrary to the behavior of bringing the weights *closer* to the axis.

FRA (9; 2). For $3W$? – '*At the end* (axis) *because the more* (weight) *there is, the closer it will have to be brought.*' Explanation: '*Because when I had 2, I was going to the middle, now I'm going to the end.* – Sure? – *Yes, sure*'. Observes his mistake. – '*Oh! It's half of that* $(1W - 2W)$.'

(2) The child seeks a direct relationship starting not from the total length but from the distance $1W - 2W$. 'It's always half the *distance from before*', i.e. by taking fractions on the remaining distance $2W$ – axis. Faced with successive failures, the child reverts to qualitative compensations knowing that distances become smaller, but without determining their size.

EDW (10; 8). 'For $3W$? – *We would need to place it in the middle* (axis) *because it's heavy.* – Why? – *Because it must be divided into 3* (confusion of limits and intervals).' Observes his failure: '*Then also in the middle as with the 2 weights*'. New failure: '*Oh, no! In the middle of half* (of what is left)'. Another failure: '*2/3 of that* (interval $2W$-axis), *no! That's wrong. Oh! We must put it in the middle* $(2W)$ *and then add this length* (1/3 of the rest)'. For the time being chance is on his side because half plus one-third of the remaining half is equal to 2/3. But when he goes on to $4W$, adding 1/4 of the rest will no longer work. For $5W$, the child reverts to approximations. – '*That always becomes smaller, more or less half of that* (half of the interval $3W - 4W$.')

(3) If the child takes into consideration the total length, he is able to discover the exact relationships by means of empirical trial and error.

TOS (10; 0) (the youngest case able to find the correct solutions): '*For 3W: 7 cm because with 3 it's 3 times heavier, thus we must move it 3 times more than the other*' (note the 3 times more). She has measured the total length which is equal to 24 cm but nevertheless continues to mix the two procedures. '*So it is 8 cm – No! With 2W we arrive at half, with 3 we can't advance just a little bit; we have to go all the way to the end* (axis).' Observes failure: '*Oh! It's just what I was told before*' and comes back to 2/3. '*For 5W: 5 cm. No, 4.5 cm. That makes 24 divided by 5. –* For 6W? – *That's 4 cm.*'

(4) The child is able to discover the correspondence between a series of weights and a series of distances, but is not able to subsequently deduce its application for any length which might be assigned to 1W.

GEO (11; 0) is able to arrive at correct solutions like Tos without being able to make the transfer when the length 1W axis is changed to 6 cm. '*For 2W? – 2 cm* (equal to 1/3) *we draw nearer, since there are less centimeters.* (Starts out from the position 1W and no longer from the total distance). – It is no longer half? – *Yes . . . 3 cm.* – And for 3W – *2.5 cm* (instead of 2 cm). – Why? – *It's half* (and no longer a third).'

(5) The child discovers the correspondence between the two ways of ordering without realizing that they involve an inverse relationship for the distances. The child says it's half for 2W, it's one-third for 3W (inverse relationships), but in fact, he takes half, two-thirds, three-fourths, etc. (direct relationships). He starts from the position 1W and not from the axis and expresses the displacement by the verb *to advance* and not *to go back*. GEO (12; 10) 'thought that for 2W it was half, for 3W, one third . . . it's like arithmetic'. But he solves the positions of W like Bern (13; 4) by 1/2, 2/3, 3/4 and explains it by advancing the weights in relation to 1W.

(d) *Discovery of the application of the law of progression.* – When the progression of the weights is 1, 2, 4, 8, etc., the child from age 11–12 can state: 'It is always half of half'. He operates by dividing the remaining distance rather than by multiplication. Dividing half by two is not identical to multiplying half by half. The child translates an inverse relationship through a direct one. Rog (12; 3) gives an example of the law: 'For 8W? – Oh! there (correct) at the 7th mark (the length is divided into 8 parts, of which the origin O is at 1W). – For 25W? – It's *easy*; we divide by 25 and place them at the 24th. – For the limit? – Oh! no . . . there is always a part left over . . . that always becomes smaller . . .'.

Thus when $W \times D = K$, the child constructs this constant relationship from the complement of the inverse of D.

NOTE

[1] This problem has already been studied from the standpoint of proportionality by Inhelder and Piaget (*The Growth of Logical Thinking from Childhood to Adolescence*, Chap. XI). We, on the other hand, have approached the problem with a somewhat different technique in that we translate the weights into units of length.

CHAPTER 13

CONCLUSION OF CHAPTERS 8 TO 12: THE GENERAL
EVOLUTION OF BEHAVIORS

In summarizing the results obtained in the five experiments with 353 subjects
aged 6 to 14, we will describe the general evolution of the behaviors
evidenced in the studies presented: *Experiment I* (rectangles), *Experiment II*
(circles), *Experiment III* (wheels), *Experiment IV* (cars), *Experiment V* (a
balance).

STAGES IN THE CONSTRUCTION OF THE NOTION
OF FUNCTION

STAGE I (up to age 7–8). – This stage is characterized by the difficulty
experienced in establishing a relation among the variables involved. We say
that a size is variable if it can successively take on values which are ordered in
a defined progression. When these sizes vary as a result of transformations
effected on the apparatus, the child has even more difficulty grasping them.
Thus it is easier for the child to designate an absolute position for a circle in
Experiment II than to predict the length of a rectangle based on its height
(*Experiment I*). The child begins designating a place not for a circle, but for a
class of circles, large ones, small ones and medium ones. The relations 'larger
than' and 'smaller than' are understood in the context of a seriation of the
elements. But in this ordering, the child does not seek to define the sizes of
the intervals because that would presuppose the establishment of a relation
between two variables. In the situation where a transformation intervenes,
this failure to establish a relation is striking. The decrease in height does not
necessarily imply even an intensive compensation of an increase in the width
(*Experiment I*). We encountered this behavior in subjects who shrink a given
figure following any transformation, i.e. when the height of the rectangle is
decreased, they make the entire rectangle smaller. We can also find this
behavior when the child does not want to go beyond the borders of a figure,
as well as in the problem of the passing of one car by another. The overtaking
(*Experiment IV*) is first attributed to the time period, and the rotational
frequency and the size of the wheel are not considered relevant.

Similarly, the non-conservation of equal distances when the wheel makes marks on the path travelled (*Experiment III*), such as the attribution of equal distances to any wheel which has made one complete turn, reveals a juxta-position of sizes which are not functionally linked. Consequently the child is not able to grasp the direction of the variation (*Experiment V*), because the variations studied presuppose the maintenance of a systematic ordering and the understanding of a causal link between the relevant factor and the effect produced.

STAGE II (ages 8–11). – During this second stage, the child tends towards the search for the establishment of a relation between the sizes involved. Starting from simple relations of intensive compensation, he tries to coordinate the results of the empirically obtained variations. Thus he begins by establishing the serial relations which lead to the law of progression of the series. The latter implies the quantification, not only of the sizes, but also of the distances of the intervals between the moments of the variations. We can further distinguish two sub-stages.

Stage IIa (ages 8–9). – The first requirement for the understanding of the variable sizes is conservation. This simple correspondence, whereby we find the same values by returning to the same state (for example: a height always corresponds to a width (*Experiment I*), a given wheel corresponds to a constant distance travelled (*Experiment III*), a suspended weight in equi-librium corresponds to a given distance from the beam (*Experiment II*)), is, in our opinion, the point of departure for any construction of functional variations. That is why we have emphasized this initial point of reference. The problem is to find out how the child establishes other points of reference during this period. Thus we see different aspects of the qualitative compen-sation of sizes. The 8–9 year-old child justifies the step-by-step compensation of the two dimensions of the rectangle without anticipating the final point of reference when one dimension is doubled and the other becomes null (*Experiment I*). (The problem of the points of reference is the same for the boundaries which we will discuss in stage III). The reciprocal compensation in this situation seemed among the simplest to achieve. Quantification was not meant to pose a problem. It is only due to the nature of the failures observed in other experimental situations that we are better able to grasp the intrinsic difficulties of this compensation which involves the comparison of the absolute or relative differences on the one hand, and the direct or inverse compensations of quantification on the other.

This quantification is not metric but is rather the establishment of a relationship of the sizes between the elements of the series as occurs in the coordination of the elements of two series. The child constructs intensive seriations from age 8 on but has difficulty with the relationships between the elements of the series up to age 10–11. This would seem to indicate a regression of behaviors, but instead points to the passage from comparison and absolute compensations to the search for relative relationships. When the child is engaged in setting up a series in which the first two terms are in a 1 to 2 ratio, he translates it as 1 plus 1 and the inverse ratio as 2 minus 1. Some say one time larger for two times larger, or half a time smaller for two times smaller. The child compares the wheels and the distances (*Experiment III*) superimposing one surface onto another, or carrying over one length onto another and he observes the ratio or the size differences; in other words, he has grasped the direct relationship. Why has the inverse not been acquired at the same time? The same applies when we introduce intermediate values, i.e. the correct solutions only appear toward age 10–11. (*Experiments III* and *IV*).

Stage IIb (age 10–11). – The 10–11 year-old child progresses insofar as he searches for the relative rather than the absolute differences in sizes. It is this possibility of establishing a relationship among the sizes which enables him to subsequently isolate the proportional relations. The following results make it possible to differentiate the two levels of stage II. When the child aged 9–10 thinks that he should place circle C_7 'closer' to C_{10}, he is estimating a surface area relationship as he would for a distance relationship, even though it is still an approximate relationship (*Experiment II*). In the seriation of sticks in the same experiment, the 10 cm stick poses a problem for the child starting from age 10, because the intervals between the sticks are considered relative to their lengths. In *Experiment IV*, the inverse relationships are only solved at age 10–11, in situations where there are only four terms to be linked: a small distance for a large wheel and a large distance for a small wheel.

Finally in *Experiment V*, the distances cannot be compared as such but only as a function of a given relative distance, starting from age 10–11.

STAGE III (from age 11–12). – The search for a law of progression of the values of a variable marks the passage from stage IIb to III. To determine the progression, two elements must be determined: the boundaries of the variation and the relationships between the successive sizes of the series. As long as the functional law is expressed qualitatively, it is enough to link the relevant variables in order to isolate the function, i.e. the height decreases as the width increases; the larger the wheel the longer the distance it travels; the heavier the weight the farther away it must be hung (towards the center); etc.

This relationship is grasped by the 8–9 year-old child. If the function is expressed by a law of progression, it is only from age 11–12 that we see the child construct the progression of values. Based on these studies, we will focus on two aspects: the limits which are to be determined, and the intervals which are to be defined.

The limits of the square are the initial square itself (or an initial side of the square) and the double thread. It is not until age 12 that the child says that all he needs to regulate all the transformations is to start from this initial square and to know the law of progression by equal and reciprocal compensation. In *Experiment II*, each stick serves in turn as a unit of length from which we must construct the intervals to place the circles. The given length becomes a base unit; this is why the child takes so long before he is able to generalize the relative position of the circles on virtual lengths. The limiting size is also used as a base unit in the balance. That distance ($1D$) may vary, but it is of necessity the base distance from which the other distances are constructed.

As regards the intervals, we note that the child may take an ordinal position for a cardinal position. In *Experiment II*, the child places the circle on the stick according to its rank in the series. The attempt to establish a relationship passes through the division of a unit into fractions, e.g., the third circle is at 3/10, which is the same as dividing the length by 10 and taking a corresponding fraction.

The relative place of each circle is thus determined first by its rank (according to the qualitative seriation) and by the fraction of the corresponding stick. C_7 is placed at 7/10 of the length of the stick. Next another condition, the absolute difference between the circle diameters, is considered. The establishment of the relationship is expressed by the formula: same difference between the diameters (circles) as between the intervals. Lastly, the child constructs the true proportional relationship: the ratio between the circles is equal to the ratio of their relative distances. The same reactions are found in the seriation of the sticks. The child at stage IIb admits that the 10 cm stick must be placed closer to S_{25} because the lengths between S_{10} and S_{25} and between S_{25} and S_{50} are different. It is not until stage III that the child is able to determine this interval through a proportional relationship (*Experiment II*).

We observe this same passage from stage IIb to stage III, when the child is asked to determine distances starting from the establishment of a relationship of the wheel sizes (*Experiment III*). This also applies to the multiplication of relations whereby the wheel size must be compensated by rotational frequency for the problem of the distances travelled to be solved (*Experiment IV*). The correct solutions require the establishment of a proportional relationship among the variables involved.

The child is very late in discovering inverse proportional relations. We even find 12–13 year-olds searching for an application of direct proportionalities (*Experiment V*). The child begins by dividing, for example, the distance D_1 by the number of weights (units of weights) to be hung. This is the same behavior found in *Experiment II* with only one difference: instead of dividing one length by the number of circles in the series, the child divides it by the number of weights to be hung. We stated that the child constructs the constant relationship $W \times D = K$ not from the inverse ratio of the distances, but by taking the complement of the inverse. The weight W is directly proportional to $D - (D/n)$ (n = the number of weights).

CONCLUSIONS

Although the idea of function is already contained in the establishment of a correspondence between two given sizes whether they be arbitrarily or causally linked, the construction of the notion of function is much more difficult. First of all, for a variation to be grasped, the order relations must be established so that the direction of the variation may be determined. Next, this variation brings about another variation whose direction must also be determined. When the relation is one which involves a causal consequence, it can help in the discovery of the intensive covariation. In other words, a functional relation which is qualitative in character can be more easily grasped. But from that point on, the constancy of the phenomenon must be established for the law of variation to be constructed. In the case where the givens of the series are present, the establishment of a correspondence among the terms would suffice. The child would not need to repeat the experiment to observe the constancy of the variation. For this constancy to be defined, the functional law must imply the search for a law of progression to quantify the variation. From this point on, a causal link no longer provides any support as was the case in the discovery of the intensive variations. The direct comparison of the concrete givens could facilitate, to a certain extent, the establishment of a relation between the elements of the series in order that the serial progression be ascertained. In any case, quantification presupposes more than an absolute comparison such as a comparison of the absolute differences; it requires the establishment of a ratio of sizes. We think that it is by starting from these ratios that the child can grasp the continuity of the variation. The deduction of the law of progression thus becomes possible. We have seen the passage from the coordination of the absolute sizes of one series and the relative sizes of another variable to proportional relations. Proportionality cannot be understood until there has been a coordination of the establishment of a double relationship or of the establishment of a relationship between two laws of progression. Thereafter, the understanding of a functional law implies the deduction of its application. This application will only be correct when the covariations are defined by the coordination of two laws of progression corresponding to two functionally linked variables.

PART III

THEORETICAL PROBLEMS

CHAPTER 14

ANALYSES TO AID IN THE EPISTEMOLOGICAL STUDY OF THE NOTION OF FUNCTION[1]

1. DESCRIPTION

The first question which comes to mind involves the legitimacy of a study which deals exclusively with the notion of function. In fact, the current standard approach considers functions as certain classes of ordered pairs, thus simply as a subset of the set of relations. This is why, on the one hand, logicians have been more concerned with developing the theory of classes and relations than with the theory of functions, and, on the other hand, why the work done in Geneva has focused on the genesis of classes and relations. Nonetheless, it must be noted that such an attitude, which is essentially reductionist, is not entirely satisfactory from an epistemological point of view. We must of course be careful each time we appeal to common sense and language in these matters, yet it seems difficult not to be sensitive to certain differences in usage between the two terms. Thus, while we can no longer speak of the function 'smaller than' (since the relation in question is not univocal to the right), neither can we speak of the function husband to wife, even in a legal context.

In fact it is easy to discover more precise differences and even to distinguish two types of such differences. The first are found at the level of mathematical application. Union, intersection and complementarity, which are fundamental relational operations, seem to be somewhat secondary operations for functions. Inversely, without claiming that the composition of two relations is a rare operation, we can nonetheless observe that it is considerably more important for functions. On the other hand, properties such as transitivity, symmetry or reflexivity which any study, however elementary, of relations cannot exclude, play a fairly minor role in the theory of functions. It can of course be very useful, for example, to know that a given function is symmetrical, in that the physical phenomena which it could represent would then exhibit interesting properties, but this point remains external to the notion itself.

The second differences are even more essential. They derive from the logical status of the two notions and are clearly manifested through the process of abstraction. $A(x, y)$ is a well-formed expression of the predicate calculus. We can thus define by abstraction the class of ordered pairs which satisfy A, i.e. the relation:

$$r = df\{(x, y) \mid A(x, y)\},$$

where $\{(x, y) \mid \ldots\}$ stands for a set operator. Let us now suppose that the calculus contains not only variables but also constants and that $B(x)$ designates a term. Thus when a constant is substituted for x in $B(x)$, the result obtained is also a constant. We will then be able to define the function by abstraction:

$$f = df\, \eta_x\, B(x),$$

where $\eta_x \ldots$ stands for a functional operator.

The fact that we use two variables x and y in one case and only one variable x in the other is totally non-essential. What matters is that we had to use the expressions A and B which are of totally different logical types. If x_1 and y_1 designate constants, rx_1y_1 is a *proposition*, while fx_1 is a *term*.

This suggests that the psychological operations involved in relations and functions might also be different. In fact we note that even though two concrete objects can be in relation to one another, (e.g. if one is to the left of the other or if it was produced previously), we cannot properly refer to an object as being a function of another. A function exists only to the degree to which a specific *property* of one is linked to a property of the other. This might lead one to think that the operations for establishing a relation might in some way be more primitive than those establishing functional relationships. However the problem is not really all that simple since we can also consider, inversely, that there are at least some relations which are only the result of the establishment of a functional relationship. It is therefore possible that the constructive aspect relies more on functions while the observational aspect relies more on relations.

So far we have only been concerned with *a priori* considerations which must be empirically verified. But since it is somewhat difficult to make pure experience bear on distinctions which are still very general and vague, a brief historical review might possibly suggest more clearly defined elements. Let us note in passing moreover, that the notion lends itself to this since, while it is impossible to pinpoint the historical genesis of notions of class, relation and even of natural number, the technical concept of function appears at a fairly well-defined moment. According to Cantor[2] the term *functio* appeared for the first time in a Leibnitzian text dated 1694. While we can use this fact as the basis for assuming that the XVIIth century was sufficiently interested in the notion as to give it a name, we are by the same token limited by it, for it is clear that even within the framework of mathematics, the concept was used well before it was baptized. This does not however prevent us from learning from history and let us note that certain mathematicians, particularly those influenced by Menger,[3] have attempted to elaborate a notion of function which, without leaving out any modern insights, would reintegrate certain historically neglected elements.

We will now proceed to sketch the development of this notion divided into four periods for convenience.

1. Antiquity and the Middle Ages

The Babylonians certainly studied what we would call functions and throughout antiquity use was made of innumerable tablets of numbers whose exact purpose sometimes eludes us but whose functional character is evident. They are in fact presented as sets of ordered pairs (x, y) and, in most cases, for a given x there corresponds only one y. For the historian, the problem is thus to understand — often to guess — how the scribe passed from the x sequence to the y sequence. It is not enough to simply discover the laws which separately engender each of the sequences. The simple term by term correspondence between x_i and the y_i is not sufficient to resolve the question since it is precisely the relational result of the construction which must be reestablished. This permits us to make the hypothesis that the idea of proportionality is very primitively linked to all functional dependences.

Let us here recall that for Aristotle all increasing functions are direct proportions,[4] that in order to interpret the Aristotelian ideas on movement, Bradwardine wrote a *Tractatus de proportionibus* and that going even further, Oresme entitled a work *De proportionibus proportionum.*[5]

In view of our experiments with children, we will emphasize three more points:

(a) First of all, the difficulty encountered by our predecessors in comparing two magnitudes of differing natures, since it was not until the XIVth century that thought was finally completely freed from the necessity to place only quantities of the same type in proportion;[6]

(b) Next, while it is quite evident and often clearly perceived that two sizes can increase or decrease together in a non-proportional manner, a fairly complex system of measurements is needed to express this. In fact we must be able to express, in one language or another, the function derived from the one being studied;

(c) Lastly, it seems that proportionality tends to eventually impose itself, like a sort of simplicity, regardless of the facts. We may recall in this respect the medieval interpretation of the slowing down of a movable object. If s is speed, f the 'force' of the object and r the resistance of the environment, we find their relationship expressed as $s = f/r$ up to the XIVth century. To pretend, as is done, that the wise men of the period were both impervious to observation and incapable of freeing themselves from the thought of the Philosopher is quite debatable and, in any case, explains nothing. The need to compare sizes of different natures and the spontaneous nature of pro-portionality (provided that we can corroborate it empirically) would at least seem to constitute a better working hypothesis.

Actually, that 'the effects are proportional to the causes' is often a commonplace among philosophers and even H. Poincaré bases himself on this point of view to characterize chance. The latter would be distinguished from causal laws precisely in that 'small differences in the causes suffice to bring about great differences in the effects'.[7] Whatever the value of the expla-nation, it must be pointed out that continuous functions whose derivative is not zero in a given interval are still preeminent and have the property that, if h is small enough,

$$f(x + h) - f(x) = h \cdot f'(x).$$

This reconfirms that for a small variation in the cause there will be a corresponding small variation in the effect, i.e. x and $f(x)$ are proportional, at least within a small interval.

2. Classical era

The classical notion of function is the result of numerous works which range, roughly, from Newton to Cauchy and Weierstrass. While it would be of interest to dedicate an in-depth study to this process, we can summarize its evolution as the progressive shift in focus from physical considerations to notions which are more properly logico-mathematical. We will linger on only one point which we consider important: at a given moment in their history, functions were elaborated by learned men whose attention was divided equally between the physical sciences and mathematics. Consequently, there resulted a totally new way of conceiving things, a way which was entirely dominated by the idea of *variables*.

While today we know that a variable can be conceived as the generic element of a set, this was not so at the beginning of the XVIIIth century when it was instead conceived as a quasi-physical magnitude which changes in time. 'I would suppose', writes Newton,[8] 'that one of the quantities proposed ... must increase by a uniform fluxion to which I would relate everything else as if it were in time'. In *La philosophie de Newton*, L. Bloch comments on this by saying that a function is 'a succession of magnitudes which are engendered in a continuous fashion by a given movement in a given time'.[9] We can further add that even Descartes, who greatly underestimated the importance of time, could not ignore the idea of variation when classifying curves into geometric and mechanical ones.[10]

The relations with which we must then deal can no longer be found solely on the plane of abstract logic. Rather, their roots go deep into concrete phenomena, phenomena which change, vary, or better yet, are covariant. This raises a new problem which, as E. Meyerson says, is that of the irrationality which accompanies change and which, with the birth of analysis, almost makes the mathematical edifice crumble. It seemed that the path which should have led to a solution would have had to pass through causality.

Let us recall the degree to which Galileo distrusted the search for causes since at the time the principle of causality seemed to be more philosophical than scientific. Nevertheless, it is clear that it does have a role to play, be it only in the postulation of the existence of a *ratio* between the sizes brought about through time. Originating at the metalinguistic level, it was not until causality emerged in the object-language that its importance was redis-covered — even though its appearance was so changed that it could no longer be recognized as such.

This came about indirectly through *constructivity*. The problem was resolved when it was clearly seen that the simple, platonic affirmation that the variable y did in fact depend on the variable x, or even that this dependence was well regulated, would not suffice, but that it was necessary to undertake the task of providing the details in terms of arithmetic operations. This is how the exigencies of construction were added to the idea of variation. In his *Comptes rendus de l'Académie des Sciences* in 1718, Jean Bernoulli writes that functions are 'quantities composed in any manner whatsoever from this variable size [x] and constants',[11] and Euler, in his *Introductio in analysin infinitorum* (1748), uses the various "manners of composition" to formulate a general classification of functions.

These brief considerations lead us to wonder if it isn't necessary to hypothesize a double genesis of function whereby it would have been jointly derived by simple abstraction with respect to its physical origin and by reflective abstraction as a mathematical concept. This would be the same as attributing to it (at least at a given point in its development) a status somewhat analogous to that of geometric objects. This is not surprising for it is known that analytic geometry played a large part in the history of functions and Clairaut, for example, wrote 'curve' where we would write 'function'.

3. The notion of application

Functions, in the classical sense, thus remained deeply embedded in a physical perspective and their introduction into the purely logico-mathematical plane required the sacrifice of the idea of variation, the abandonment of time, and consequently, of causality.

It is remarkable that Euler himself realized this for, in the above-cited *Introductio*, he conceives of a variable in the modern sense as a simple generic element: *'Quantitas variabilis est quantitas indeterminata seu universalis, quae omnes omnino valores determinatos in se complectitur'.*[12]

Today however we know that in order to fruitfully develop the notion of application, we must first elaborate a satisfactory theory of sets. From the psychological point of view, there is enough information available on the genesis of classes and relations for us to be able to dispense with having to go back over it, at least directly. History, however, always suggests an indirect return.

It is in effect standard practice to distinguish two procedures for determining a set. The first consists of listing the elements, is purely extensional and is only applicable to finite sets. The second, starting from a defining property, consists of constituting the set by abstraction. There is also a third procedure whose importance has been pointed out by logicians and which can be found mid-way between the first two. It involves the recursive generation of sets which gives rise to the theory of recursive functions. We believe that therein lies a point of fundamental importance. It is first of all important for epistemology since recursivity expresses the constructive element introduced above. Next, of course, for logic where recursivity cannot for all practical purposes be distinguished from the idea of a formal system. And finally for psychology, insofar as it places genuine operations, i.e. composable and reversible actions still *to be effected*, at the core of its explanations. In this third method, then, the subject is still present to a degree since he is active, constituting the extension which he needs through the intension given him or which he himself has obtained.

It would thus appear that the idea of application comprises more than a simple class of ordered pairs, even more than a univocal correspondence to the right. We can find contained in this idea all the operations whether mathematical or otherwise, which engender correspondences. The concept of an application as a relation is certainly correct, but incomplete, because as such it only takes into account one result, leaving aside the mechanisms which engender it.

4. Morphisms

Until now we have spoken of functions and of applications in the plural, thereby indicating that we considered them isolated individuals, which is certainly legitimate. The 'square' function and the 'sine' function have very distinct properties, as does the number 6, which is even and divisible by three and differs fundamentally from the number 7, which is odd and prime. An atomistic study of concrete or abstract objects can provide a wealth of information and studies of this type abound during the early stages of the various sciences. Nevertheless, as the example of the numbers demonstrates sufficiently, it is still much more fruitful to ask oneself about the entire category of the beings under consideration, thinking more about what unites them into families than about what separates them into individuals.

This is the perspective that Brandt subscribed to as early as 1926[13] and it is also how Menger proceeded in his diverse studies on functions.[14] Thus the study of the category of morphisms, in the technical sense of the terms, corresponds to a new period in the history of the notion with which we are here concerned.

This contemporary viewpoint provides us with three areas for psychological analysis:

(a) It makes it possible to focus attention on whatever interest there might be in making experience bear on the composition of functions as much as on the elaboration of a particular function. Let us not forget that the analogy with the numbers is probably somewhat misleading. The young child knows very early how to recite 1, 2, 3, 4. It is easy to provide him with physical situations which properly illustrate the operation of addition, and the numbers are thus finally ordered by the operation which engenders them. None of this is true for functions. The child does not learn a list of elementary functions; it is fairly difficult to 'give' two functions and their composition; and finally, there is nothing very natural about ordering one function before another. Nevertheless, it falls precisely within the realm of genetic epistemology to try to look at what is an important mathematical fact in light of the spontaneous development of thought, and the difficulty of this task does not take away from its validity.

(b) It emphasizes the importance of distinguishing monomorphisms from epimorphisms. The distinction will no doubt appear secondary if we only study functions by means of elementary physical devices. In fact, the law of the stretching of a spring, that of vibrating cords and so many others are always exemplified as one-to-one correspondences. One could thus object that this is a question of highly elaborated distinctions of use only to mathematicians.

However, without underestimating the practical difficulties of concretization, we do not think that these objections should restrict us. In the final analysis, the mono-epi distinction rests on a simple operatory criterion.[15] The difference is made concrete for the child when he is presented with objects to be put into boxes. He must then be able to distinguish two situations:

(1) *No more than one* object in each box (injection which corresponds to the mono characteristic);

(2) *At least one* object in each box (surjection which corresponds to the epi characteristic).

(c) Finally, it leads to an even more fundamental question, i.e. on the precedence to be ascribed to the origins of classes. Given the degree to which a class is never considered a class by the child unless he has somehow constructed it, we might think that the operations which he will put into play are somehow prior. In point of fact, such logical distinctions are only used for the sake of convenience and we must realize that classes and operations sustain among themselves relationships of an inherently dialectic nature. But, precisely in this case the language of morphisms is a powerful tool.

It should therefore be possible at least in principle to elaborate the theory of categories with only morphisms, without associating 'objects' to them. But we know that in cases where objects are sets, injections correspond to morphisms and surjections to epimorphisms. And since any application defines a quotient-set on its source-set and a subset on its target-set, we see that three of the basic notions of logic are related to morphisms, i.e. the notions of class, of hierarchical inclusion and of quotient-set or equivalence class.

Returning to our original problem on the genesis of the notion of function, it would seem that we can make three global *a priori* hypotheses:

(1) The notion of function could result from the coordination of certain other notions elaborated more or less independently of it. Consequently we could expect to find in it the notion of relation, very likely also that of the product of two relations as well as that of number. We would also expect to see causality and causal implications play a certain role.

(2) Inversely, we could posit an autonomous genesis which would eventually derive from instrumental actions, – if causality truly plays a role therein – pass through groupings and thereupon become differentiated. Thus groupings would be the common source of the natural numbers with their own structure, of propositional logic with its Boolean algebraic structure, and of functional algebra, i.e. Brandt's groupoid, and Menger's hypergroup.

However it must be noted that, given the present state of formalization of child behavior, there is not much likelihood that we will find models which are subtle enough as to differentiate between the two hypotheses for the youngest children. Furthermore, it seems unlikely that the emergence of one behavior subsequent to another would provide an argument in favor of one hypothesis and not of the other. In fact, the difficulty of coordinating already elaborated notions leads to delays which are comparable to those resulting from the construction of a specific notion.

(3) The last hypothesis is not strictly speaking an intermediate one, although it does borrow certain characteristics from each. It was proposed by Piaget and by Bresson and is essentially based on the distinction of two types of functions: *structuring functions* and *structured functions.*

Structuring functions, would be found at the beginning of the genesis: they would, at least in the sense that they are engendered by actions, have properties which are quite akin to operations and would probably satisfy hypothesis (2). Structured functions, would be found, so to speak, at the point of arrival. As they are in effect composed of diverse elements, and structured by the operations themselves, they would fulfill some of the conditions of hypothesis (1).

Hypothesis (3) sacrifices nothing which is essential to the first two. It even offers the advantage of allowing broader empirical input and could, for example, lead to showing that the 'logic' of structuring functions is only a part of the 'logic' of structured functions (see Piaget's conclusions in Chapter 15).

2. STRUCTURING FUNCTIONS AND CLASSIFICATION

If there truly exists some psychological reality which can confirm the hypothesis of structuring functions, we believe it must be sought in the simplest classificatory activities. In fact, the standard procedures of classification by similarity and difference derive from behaviors which are quite well developed. At any rate, the awareness of the procedures involved must emerge fairly late. We may recall that it was not until Porphyry (234–306) isolated the *quinque voces* that a grouping structure was completely established. There are two other arguments which support the thesis that this type of classification is highly evolved.

The first derives from experimental psychology but we will not go into it here for it is quite well known. Piaget and Inhelder, have clearly shown[16] that 'classes' are preceded by simple figural and non-figural 'collections' into which the objects are united not because they possess any common property but because together they satisfy one criterion or another, e.g., they may represent a familiar object (house, train, etc.) or suit a situation imagined by the child (shepherd and sheep resting in the shade, etc.).

The second derives from an *a priori* logical analysis. Any partition of a given set E — and this is precisely the problem faced by the child — leads to the establishment of equivalence classes. Thus what must be understood is why objects which in themselves often have as many differences as similarities, end up 'being put together'. The answer 'because they share such and such a property' only appears to be simple. We must understand why the property which they have in common finally supersedes those which would tend to separate them. It is evident that, outside of the laboratory, it is the intended action which underlies the preference for one property over another. When the object is to drive in a nail, a hammer, a hatchet, pliers or even a stone seem to be equivalent. We can thus hypothesize that it is the projected action itself which increases the value of certain properties to the detriment of others and that it might be possible to isolate a certain number of preliminary operators.

We must naturally specify their nature. The language of combinatory logic seems highly appropriate in this regard. Louis Frey has already used it in several studies[17] and Apostel has gone as far as to claim that he was the only one able to adequately report on the behaviors of children.[18] The following schematic example should serve to sufficiently refine our thoughts on the subject.

Given three objects *a, b* and *c*, let the problem be to effect the partition (*ab.c*). We introduce a classificatory operator γ which can correspond to several simple concrete actions, depending on the case. Next we introduce a union operator ρ which unites the objects 'classed' by γ. Thus the formal problem is to find an operator X such that we have the following rewriting rule:

$$X\rho\gamma abc \rightarrow \rho(\gamma ab)(\gamma c).$$

Since this is an imaginary example, the detailed form of X matters little here. It suffices that we know we can find it and that it contains, in particular, the operators I, W, C and B. As we recall I is an identifier and W is a repeater. It would indeed seem to be the case that any classificatory procedure must necessarily begin by 'identifying' the objects on which it bears and that it could not develop without 'repeating' certain sub-procedures. As for the operator C, it can be defined by the rule:

$$C\phi\alpha\beta \rightarrow \phi\beta\alpha.$$

According to Apostel, it would represent seriation but together with Piaget, we would rather consider it as retroaction without excluding, as a refinement of the technique, the introduction of the operators C_m of differing powers. Finally, still according to Apostel, there would be the operator B:

$$B\phi\alpha\beta \rightarrow \phi(\alpha\beta)$$

which represents the actual classificatory element.

Of course we can ask ourselves if the operators used here, which are taken as primary by logicians for various reasons, are also the most elementary from the psychological standpoint. This is no doubt an important question and we could possibly imagine a series of experiments specially conceived to answer it. In any case, it may be less fundamental than it seems. It is clear that the combinatory language is above all an analytical instrument which can be most useful even when its elements do not correspond term-by-term to the operations which can be discerned in children.

Whatever the case, it is methodologically possible to provide for two species of equivalences, even if one is considered to be no more than an artificial refinement of the other, or if on the contrary, the least developed one appears as such only for observational reasons. The first would include *direct equivalences*, i.e. equivalences based on the relations of coproperties of objects, such as colors, further reinforced by the linguistic practices of the child. The reflexivity, symmetry and transitivity of such relations would then not be truly elaborated but simply observed.

By contrast, the second would comprise *indirect equivalences*, i.e. the equivalences constructed with the help of elementary structuring functions. Limited and local operators, α, β ... , would provide reasons for 'putting together' certain groups of elements of a given set E. Logical and general operators of the type cited above would lead to the coordination of these local operators, the result X being nothing more than a structured function. This function which is then conceived as an application of a given set into the constructed 'boxes' would lead to a partition of E, i.e. to the quotient-set by the equivalence relation engendered by X.

Although the procedure may appear complicated, it nonetheless offers some appreciable advantages. Irrespective of the fact that it has proved itself extremely useful in algebra, it also enables us to direct experimental attention onto very important behaviors. Furthermore, it would address itself to a frequently made observation, i.e. that any 'practical' classification – in contrast to a scientific and systematic taxonomy – does not immediately derive from a conscious and explicit equivalence relation. Finally, it would explain one of the formal characteristics typical of functions with respect to relations, i.e. univocalness to the right. It would be nothing more than the expression of the operatory and constructive aspect of classification.

In the light of the work already done by Piaget and his collaborators, we could conceive of a double classificatory level. On the one hand, we would find a classification based on successive hierarchical inclusions. The fundamental relation would be inclusion and the basic operations would be union and its inverse. We know that such procedures lead to a logical structure of Boolean lattices. On the other hand, we would find a classification which would proceed by local constructions. Here the fundamental relation would be that of restructuring (corresponding to re-writing rules) with the operations differing from one group of elements to another.

Finally, it is interesting to note that while a single partition of a given set does not have a predetermined structure, the set of all the possible partitions of a finite set E is also a lattice. The fact that the set of the subsets and that of the partitions both have a lattice structure is encouraging, but at the same time, it poses the question of their coordination. It would no doubt be premature to attempt to make any hypotheses here, although it should not be impossible to study the question once we have at our disposal a sufficient number of empirical facts about structuring functions.

3. STRUCTURED FUNCTIONS AND PROPORTIONS

There are two principal ways of logically defining an ordered pair. The first makes use of propositional forms and defines an ordered pair as follows:

$$(a, b) = df\{xy \mid x \in \{a\} \land y \in \{b\}\}$$

or equivalently:

(1) $$(a, b) = df\{xy \mid x = a \land y = b\}.$$

The sign '=' here represents logical identity.

The second way is more direct and defines an ordered pair as follows:

(2) $(a, b) = df\{\{a\}, \{a, b\}\}$.

Since both definitions lead to the same formal rules, there is no disadvantage in writing (a, b) both times for the definiendum. We nevertheless think that the naive ideas underlying the two definitions are distinct and it is from this point that we will start.

What is most important in definition (1) is the presence of an identity relation, thanks to which it is possible to establish an equality between ordered pairs:

(3) $(a, b) = (c, d) \Leftrightarrow a = c \wedge b = d$.

On the other hand, identity is an abstract concept which is only the limit of all the possible similarities between two objects. It follows that in all applications, we only really deal with partial identities, i.e. equivalence relations under such and such a property.

Let ϕ be an equivalence relation, then (1) can be naturally generalized as follows:

(4) $[a, b]_\phi = df\{xy \mid x\phi a \wedge y\phi b\}$.

Since ϕ is reflexive, symmetrical and transitive, we would again have

(5) $[a, b]_\phi \approx [c, d]_\phi \Leftrightarrow a\phi c \wedge b\phi d$.

There remains for us to understand the significance of these formal equations. Let us suppose that we are given a set E whose elements are the lower case Latin letters. As we saw in the preceding paragraph, the introduction of an equivalence relation on E comes to be the same as constructing a partition of E. Relation (5) only expresses the fact that two elements which are equivalent under a given property can be substituted for each other.

Let us now turn to definition (2) which reduces the notion of an ordered pair to only two ideas: there exist two elements (for logical reasons, the classes $\{a\}$ and $\{a, b\}$) and furthermore, there is a first element, followed by a second. These are the very ideas which Piaget showed to be constitutive of the notion of number, i.e. cardinality and ordinality, and which are synthesized in natural, thus finite, numbers.

We will not focus our attention on this, however, but rather suggest that the remaining order is only the vestige of a prior construction which has been abstracted. In other words, it seems useful to generalize (2) by rewriting it as:

(6) $(a, b) = df \{\{a\}, \{a, \tau(a)\}\}$ with $\tau(a) = df \, b$,

where τ is a given transformation operator.

The simultaneous consideration of generalizations (4) and (6) leads to what we could call a notion of *preproportionality*. For example, let a set E be composed of objects of different colors. We saw above that there was reason to posit an entire set of elementary operations or structuring functions, which would lead to a partition of E into a certain number of equivalence classes relative to the relation ϕ (in this case, having the same color). If $a, b \in E$, a and b will belong to the equivalence classes C_a C_b. Let us assume that they are distinct. Then definition (4) gives:

(7) $[a, b]_\phi = df \, C_a \times C_b$.

where '$\times$' designates the Cartesian product. As for relation (5) we can write it in a more descriptive form as follows:

(8) $\dfrac{a}{b} \approx \dfrac{c}{d} \Leftrightarrow (a, b), (c, d) \in C_a \times C_b$.

It then follows that we also have:

(9) $\dfrac{a}{b} \approx \dfrac{c}{d} \Rightarrow \dfrac{a}{d} \approx \dfrac{c}{b}$.

Although this type of preproportion is still fairly elementary, it nonetheless represents a first enrichment of simple partitions in the sense that it expresses the invariant which permits the passage from class C_a to class C_b. The introduction of the transformation τ in true logical proportions such as those studied by Piaget[19] can only reinforce its importance. We note:

$$\dfrac{a}{b} \doteq \dfrac{c}{d}.$$

One of their essential properties is that they lead to the relation:

(10) $\dfrac{a}{b} \doteq \dfrac{c}{d} \Rightarrow \dfrac{a}{c} \doteq \dfrac{b}{d}$.

We might of course wonder why there is as yet no such implication at the level of preproportions. The answer is fairly simple and relies specifically on the presence of τ as opposed to ϕ. We cannot introduce a non-identical transformation within an equivalence class without contradiction, i.e. we cannot simultaneously state that two elements, a and c for example, are indistinguishable because both belong to class C_a but are nevertheless distinct, since one results from the other by τ.

In contrast, we can ask ourselves whether it might not be convenient to introduce an intermediary between the pairs linked by '$\approx$' and those linked by '$\doteq$' to account for the well-known analogies of the type 'Paris is to France what Rome is to Italy'. We would have to start with two distinct sets E (capitals) and F (countries). The problem, however, is complicated in two ways. On the one hand, it is evident that we could not merely consider the product $E \times F$ – the relation 'Paris is to Italy what Rome is to France' makes no sense, – on the other hand, we cannot, without being arbitrary, apply only relation (10) and exclude the others. It thus seems that the solution should be sought in the subset of $E \times F$ constituted by the one-to-one and onto correspondence (bijection) 'Capitals $\rightarrow$ Countries'. In this way we once again return to structuring functions.

Finally, let us examine the sense in which true proportions are more advanced than preproportions. It suffices to recall the definition given by Piaget:

$$(11) \qquad \frac{a}{b} = \frac{c}{d} = df \begin{pmatrix} a \cup d = b \cup c \\ a \cap d = b \cap c \end{pmatrix}.$$

We immediately see that the purely relational point of view has been enriched by the presence of two operations, '$\cup$' and '$\cap$'. This raises two questions.

First, under what condition can we compose among themselves elements which appear in the 'numerators' and in the 'denominators,' respectively? It is clear that it can only be done by renouncing any species distinction among them, in other words by using the least fine limit-partition of E possible, in which the elements are united by the simple fact that they belong to E. It then follows, as we had foreseen, that logical proportions now only take into consideration the sole transformational invariant τ. They thus abstract out certain nuances, but they nevertheless enrich the situation. This leads us to the second question: what sort of new structure do the operations '$\cup$' and '$\cap$' bring to E?

Due to the lack of experimental data, we cannot provide a well-founded answer, and we will limit ourselves to making three remarks.

(1) Piaget developed the theory of logical proportions within the

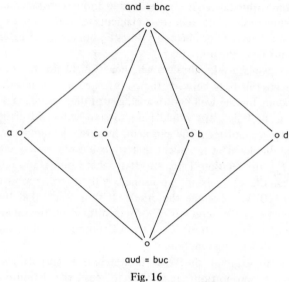

Fig. 16

framework of Boolean algebra from which it derives all of its significance. However it could be of some interest if we could find out, at least for some of the important properties and in particular the one expressed by relation (10), whether they are not already contained in weaker structures. We see, for example, that the lattice presented above (Figure 16) preserves the double equality (11). This is a non-distributive lattice as we can see by calculating $a \cap (c \cup d)$ and $(a \cap c) \cup (a \cap d)$, but it is nonetheless modal and corresponds to certain exigencies of indeterminacy of quantum physics.[20] Furthermore, it addresses, as we hope to show, certain fundamental situations which spontaneous thinking must confront, such as when several types of oppositions are involved. Let a denote beautiful, b: not ugly, c: not beautiful, d: ugly. Thus $a \cap d = \wedge = b \cap c$ and also $a \cup d = \wedge = b \cup c$.

(2) It is very likely that the problem of proportions is also closely linked to that of negation which already appears at the level of preproportions. It seems natural to suppose that the most primitive partitions will be dichotomous, especially since so far only pairs rather than n-tuples are involved. Under these conditions, it is quite natural for the complement of C_a to become indistinct from C_b and for the complement of C_b to become indistinct from C_a. Furthermore, if the two poles of the above lattice are $\wedge$ (upper pole) and $\vee$ (lower pole) respectively, the lattice is ortho-

complemented. This means for example, that not only is d a complement of a, but b and c are complements also. We thus have every reason to think that the Boolean uniqueness of the complement is, once more, the expression of a limiting-case. Finally, if we posit $a = c$ and $b = d$, we obtain a Boolean lattice:

$$\frac{a}{b} \doteq \frac{\overline{b}}{\overline{a}},$$

whose epistemological impact has been shown by Piaget in the works cited.

(3) Lastly, as soon as we have a lattice structure, and with no other condition, we have:

$$\frac{a}{a \cap b} = \frac{a \cup b}{b},$$

and thus, specifically:

$$a \cup b = (a \cap b) \cup (a \cup b).$$

It may be of interest to observe the formal analogy between this last expression and the fundamental condition for any probability function on a Boolean algebra:

$$pr\,(a) + pr\,(b) = pr\,(a \cap b) + pr\,(a \cup b).$$

Of course, this remark cannot take on any meaning until after values have been substituted. Nonetheless, it might be useful, in view of recent studies, to pay close attention to it.

All of the preceding only relates to the logical aspect of proportionality and there still remains for us to understand how these considerations can concern concrete physical phenomena. We have already suggested that this might take place through the intermediary of a concept of size, which is in fact prepared by preproportions since they presuppose the partition of a given *set E into a certain number of equivalence classes C_i. Furthermore — and this is the fundamental point — they permit the passage from one class to another through a constant transformation.

Here again we find the problem of the order which results directly from actions and which is as it were irreversible, and its passage to hierarchical inclusions. But this point presents no theoretical difficulty.

For example, let the sets $E = \{a_1, a_2, b_1, b_2, b_3, c_1, c_2\}$ and $F = \{a, b, c\}$ and let $f : E \rightarrow F$ which leads to the partition of E into three classes: $\alpha = \{a_1, a_2\}$ $\beta = \{b_1, b_2, b_3,\}$ and $\gamma = \{c_1, c_2\}$.

(1) α corresponds to a, or put another way, the image of a_1 and a_2 is a thus $\{a\} \subseteq F$. Likewise β and γ lead to $\{b\} \subseteq F$ and $\{c\} \subseteq F$.

(2) α, β and γ are disjoint (partition of E). As a result, the operation of exclusive union '+' can apply to them. We can thus have:

$\alpha + \beta$ which engenders $\{a, b\} \subseteq F$ such that $\{a\} \subseteq \{a, b\}$ and $\{b\} \subseteq \{a, b\}$.

$\beta + \gamma$ which engenders $\{a, c\} \subseteq F$ such that $\{a\} \subseteq \{a, c\}$ and $\{c\} \subseteq \{a, c\}$, and so on.

In short, the operation '+' on equivalence classes leads to engendering all of the possible hierarchical inclusions of the image-classes (the lattice of the subsets of F).

(3) (Reversibility) $\{a\}$ is associated with class α, $\{b\}$ with class β, etc. and $\{a, b\}$, for example with class $\alpha + \beta$ or with $\{a_1, a_2, b_1, b_2, b_3\}$. Since $\{a\} \subseteq \{a, b\}$, i.e. since F is hierarchically included, E is also hierarchically included. Here: $\{a_1, a_2\} \subseteq \{a_1, a_2, b_1, b_2, b_3\}$. Furthermore, the procedure is dichotomous, since:

$$(\alpha + \beta) - \alpha = \beta = \{a_1, a_2, b_1, b_2, b_3\} - \{a_1, a_2\}$$

We thus see that preproportions lead to knowing that for every object $x \in E$ there is a class C_x (the image of x by canonical application of E onto its quotient-set) and that these classes are ordered among themselves. Therein lie both the ideas of size and of measurement, although not yet numerical.

Now let us suppose that, as in the situations studied by Inhelder and Piaget,[21] the subjects had the task of first discovering, then of stating or of representing, certain physical laws. In the simplest case, the problem consists in discovering two equivalence properties ϕ and ψ which will lead to two canonical applications F and G. In other words, the child must order the objects according to two sizes. It is likely that, from the viewpoint of the subject, the problem is not exactly the same when the elements of E are simultaneously presented in space as when these elements are the successive states of the same object. However, for the purposes of our analysis, the difference is of little import, given the fundamental fact that a solution is only possible if the two partitions coincide. This means that in order for a law to be isolated, those properties which are somehow physically related to each other must first be discovered. This is why, if we have a certain number of pendulums of varied lengths and colors, the chances of arriving at a law of isochronism are very slight if we classify them by color. This is not a spurious observation but rather one of the mainstays of causality. Any variation in the length of a pendulum has as an *effect* a variation in its period, while the variations in its color generally have no effect. From this perspective, we thus see that although causality might well have been historically linked with the genesis of functions, it only really comes into its own as we seek to apply certain logico-mathematical constructions to the physical world. This is not surprising if together with Bunge[22] we accept that causality is, above all, a type of determination. Constructions on the formal plane are applicable by virtue of a total range of possibilities, and two sizes can *a priori* engender identical partitions. However, in reality, they would only do so if there existed a physical link of causality between them.

Let '$\rightarrow$' be the order relation between the equivalence classes F_i, on the one hand, and G_i, on the other. There are three possibilities:

(D) $F_i \rightarrow F_j \Rightarrow G_i \rightarrow G_j$

(I) $F_i \rightarrow F_j \Rightarrow G_j \rightarrow G_i$

(C) $F_i \rightarrow F_j \Rightarrow G_i \leftrightarrow G_j.$

We will call these three relations *functional laws; direct, inverse* and *constant*, respectively. They express certain types of dependences between the sizes considered and they implicitly contain the ideas of variable, variation and covariation which we encountered in the historical review. Furthermore, as Piaget has shown,[23] all asymmetrical relations, such as the relation '$\rightarrow$', express ordered differences. If we recall that logical proportions also enjoy the property:

$$\frac{a}{b} \doteq \frac{c}{d} \Rightarrow \frac{a}{b} \doteq \frac{c-a}{d-b},$$

where $x - y = df \, x \cap \bar{y}$, we can see that (D) and (I) can be expressed by the following formulas:

$$\frac{F_j - F_i}{G_j - G_i} \doteq \frac{F_h - F_i}{G_h - G_i} \quad \text{and} \quad \frac{F_i - F_j}{G_j - G_i} \doteq \frac{F_i - F_h}{G_h - G_i}.$$

Note that while the classes indexed i can already be interpreted as units in this type of non-numerical measurement, the interpretation in terms of proportions of law (I) does not yet correspond to the 'inversely proportional sizes' of arithmetic. Rather it only expresses the fact that one of the sizes decreases when the other increases. It therefore follows that the significant change which takes place when this is raised to the arithmetical level is the introduction of the operation of multiplication which will, in effect, make it possible to replace F_j and F_k in the above proportions by multiples of F_i and to proceed likewise with the G's.

Although this remark may appear trivial, it can nonetheless help us to emphasize two difficulties which the child might encounter. The first is on the plane of action itself. If it is true that certain links of a causal nature are involved in this type of problem and if it is also true that the child conceives of a qualitative ordering as a sequence of differences, it is very probable that young subjects will begin by reducing the size relationships to simple comparisons of differences and will confuse the rank (ordinal aspect) with the addition of 1 (cardinal aspect). The other difficulty, which might further reinforce the first, is logico-mathematical in nature. The natural numbers are nothing more than the successors of zero obtained through the operation of adding one. Thus they are, so to speak, the concrete manifestations of addition. There is something contrived about effecting any new operation such as multiplication on them and it would be interesting to study the genesis of this in detail. It is true that the product of two numbers can be more or less reduced to the intersection of two classes. But we also know the precautions with which such logical reductions must be interpreted.[23] Let us further note that, even from the formal standpoint, the introduction of multiplication is laden with consequences. The methods required to prove the consistency of a system with multiplication must be essentially more powerful than those necessary to prove the consistency of a system which contains only addition.[24]

4. ELEMENTARY RELATIONS-OPERATIONS

Piaget recently pointed out the continued indecision even among mathematicians, as to the distinctions between operations and functions.[25] In fact the preceding discussion could account for this indecision if, as we have suggested, the constructive aspect is the fundamental element. However, the relationship between intension and extension still remains to be examined once more. What we know about this relationship leads us to accept these three facts:

1. The two notions are so deeply linked that it is impossible to conceive of one without also thinking of the other, not only *in abstracto* and philosophically, but in any situation, however concrete it may be;

2. If it were nonetheless necessary to ascribe a sort of priority to one of them, we would most likely lean towards intension. The degree to which it depends on language is as yet unresolved. We can only remark that any

definition made solely on the plane of discourse, without recourse to deno-
tations, refers to the properties of objects and not to the objects themselves;

   3. By contrast, almost the whole of logic and mathematics proceeds by
extension. Let us begin by noting that, when viewed in the light of point 1,
facts 2 and 3 are in no way paradoxical. If intension and extension are each
truly the counterpart of the other, nothing prevents us from attributing a
greater significance to one in certain tasks and to the other in certain others.
This leads us to ask, on the one hand, whether there does not exist at least
one mathematical domain where the two notions can effectively substitute
each other, and on the other hand, whether we cannot find a notion where
they would appear clearly simultaneous. We think that morphisms and
ordered pairs fulfill this twofold condition while providing models of concrete activities.

   It would be difficult to support the statement: 'in the beginning there was
Class'. It is for this reason that, together with Apostel,[25] we have searched
for operatory criteria to decide when it is legitimate to say that a subject
possesses a class and when he doesn't. We will not review Apostel's
observations and definitions here and we will limit ourselves to starting from
the notion of collection. A collection 'contains' objects, but it is distinguished
from a class or a set in that it remains indeterminate in two ways. On the one
hand, there is no reason for a given object to belong or not to belong to it. On
the other hand, and this is a consequence of the preceding, it does not
constitute a totality closed on itself (Figure 17, a).

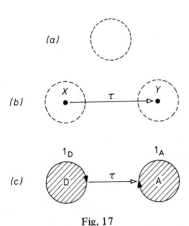

Fig. 17

What we know about the activities of the subject leads us to hypothesize an action τ which can be separately applied to objects x leading to objects $y = \tau(x)$ (Figure 17, b). The operator τ would then have the effect, as we will specify below, of constituting the x as well as the y objects into classes or sets (the source-set D and the target-set A) and of doing so simultaneously in intension and in extension (Figure 17, c). We know that in a categorical structure, 'unit arrows' 1_D and 1_A are associated with any arrow τ and furthermore, that there is a one-to-one correspondence between any arrow 1_x and an 'object' X.

It is evident that such a formalism, or, if we prefer, such a language, supports numerous interpretations. We propose to introduce the following one here: to the 'unit arrows' there corresponds the intension of the sets and to the 'objects' their extension. This means that 1_D, for example, provides the reason whereby the x's will in fact be counted as elements of D. Said another way, and in accordance with what Piaget has expressed on many occasions,[26] 1_D appears as a 'co-something relation' and D is nothing other than the corresponding equivalence class.

One could doubt whether this language contributes anything new to our knowledge of classes and relations since a language is by itself incapable of adding to knowledge. This does not however exclude the fact that a well-chosen language can suggest research which can be fruitful. In our opinion, such a language at least offers the three following advantages:

(1) It accounts for the three facts indicated above: indissociability of intension and extension, the likely priority of the former, and the fact that manipulation is only possible with the latter;

(2) It makes it possible to speak, with as much precision as desired, of applications as such, i.e. it can validate research which does not presuppose the class-function logical order;

(3) Finally, it leads to the attribution of a central role to ordered pairs, in terms of the notions of operation, class, relation and function.

Here we think it is interesting to point out that Peirce was no doubt the first to draw attention to the fundamental role of ordered pairs, which he in fact called 'elementary relations'.[27] He showed, in particular, that if we consider four dyadic relations such as 'colleague of', 'teacher of', 'student of', and 'classmate of', only certain compositions are possible. This led him to the idea, which we consider fundamental to the study of the genesis of intelligence, of an operation which is not everywhere defined. Actually, this is one of the central ideas of the structure of a hypergroup, as introduced by Menger. Peirce, basing himself on the studies conducted by his father on quaternions, remarked with surprise on the analogy between the table of composition of these four relations and that given by B. Peirce for the quaternions.

Today it seems possible for us to carry this idea a little further. Instead of asking by what means and under what conditions order is introduced into the class $\{a, b\}$ to give rise to the ordered pair (a, b), we can start from the pair itself. The question of the disappearance of the order is easily resolved by the constitution of the domain and of the codomain (the above mentioned sets D and A). On the other hand, the order $a \to b$ has no meaning without the inverse order $b \leftarrow a$. That (a, b) and (b, a) are two aspects of the same reality is inherent in the idea of order, in contrast to that of class. Finally, if in formula (2) of the preceding paragraph, we let $a = b$ we get:

$$(a, a) = \{\{a\}, \{a, a\}\} = \{\{a\}, \{a\}\} = \{\{a\}\}\, a$$

Ignoring the fact that $\{\{a\}\}$ is a class of a class — which does not seem essential outside of a logical formalism which in the end is fairly specific — we observe that (a, a) only refers to a as (b, b) only refers to b. We are thus naturally led to consider the set:

$$M = df\ (a, a), (a, b), (b, a), (b, b)\ .$$

We thus see that, if x, y, $z \in M$:

(1) $(xy)z$ is defined if and only if $x(yz)$ is defined and then $(xy)z = x(yz) = df\, xyz$;

(2) If xy and yz are defined, then xyz is defined;

(3) For any $x \in M$, there exists a u and a $u' \in M$ such that ux and xu' are defined and $ux = x$ and $xu' = x$.

Since formally, any set M which enjoys properties (1) – (3) is an *abstract category*,[28] we can conclude that the fairly natural set of four pairs mentioned above is a category.

It might appear somewhat impertinent to speak of 'categories' in the trivial case we have chosen. Nonetheless, it is correct and follows a direction which has proved fruitful. It has been some time since Piaget made the fundamental observation that the structures elaborated by the child in the course of his development were none other than those considered at the time by the early Bourbaki as the three mother-structures of mathematics. Therefore today, when mathematicians are seeking to delve more deeply into the foundations of the mathematical edifice by isolating the notion of category, it would be another step in the same direction if this same notion could be discovered in children through psychological experimentation.

5. PROVISIONAL CONCLUSIONS

We cannot propose conclusions without submitting the preceding to factual verification. However, it might be useful to suggest the following ideas.

First of all, if in accordance with all that we know, we place first the actions, then the operations of the subject at the center of the genesis of intelligence, we are led to ascribe a very special significance to operators. These seem to have a particularly simple manifestation in ordered pairs, which thus appear at the root of classes, relations and structuring functions.

Furthermore, next to the classifications made by successive hierarchical inclusions, there could exist classifications by partitions. It would then follow that the fundamental coordination of 'some' and of 'all' could still be expressed in terms of the duality of injection and surjection.

It is also possible — and this would permit us to account for the precocious successes in certain tasks — that the structuring functions are, at a given moment in the genesis, a sort of concrete substitute of the future combinatory operations.

Lastly, we cannot exclude the possibility that class is no more the most basic psychological reality than set theory is the last word in the foundation of mathematics.

NOTES

[1] By Jean-Blaise Grize.
[2] Cantor (1901), III, 215.
[3] See, for example, Menger (1959) which has a bibliography on other works by the author on the subject.
[4] Milhaud (1906), 112–117.
[5] Grant (1960).
[6] Crombie (1961).
[7] Poincaré (1920), 79.
[8] Newton (1740), LIX.
[9] Bloch (1908), 79.
[10] *Géométrie*, book II.
[11] Cited by Cantor (1901), III, 457.
[12] 'A variable quantity is an indeterminate or universal quantity which totally encompasses in itself all of the determinate values', Euler (1913), I, 2.
[13] Brandt (1927).
[14] Menger (1959).
[15] See for example Guilbaud (1961–1962).
[16] Inhelder-Piaget (1964).
[17] Frey, 1967.
[18] Apostel (1966).

[19] Piaget (1952); Inhelder-Piaget (1958).
[20] Mittelstaedt (1963).
[21] Inhelder-Piaget (1958).
[22] Bunge (1959).
[23] Piaget (1949).
[24] See for example the study by Papert (1960).
[25] Hilbert-Bernays (1934), I, 7, d.
[26] Piaget (1966), 20–21.
[27] Apostel (1959).
[28] For example in Piaget (1949).
[29] Peirce (1933), III, in particular *Paper III* of 1870 and the 1883 appendix.
[30] Mac Lane (1963).

BIBLIOGRAPHY

Apostel, L., 'Logique et apprentissage', *Études d'épistémologie génétique*, VIII, Paris, Presses Universitaires de France, 1959, 1–138.

Apostel, L., 'Psychogenèse et logiques non classiques', *Psychologie et épistémologie génétiques*, Paris, Dunod, 1966, 95–106.

Bloch, L., *La philosophie de Newton*, Paris, Alcan, 1908.

Brandt, H., 'Über eine Verallgemeinerung des Gruppenbegriffes', *Math. Annalen* 96, (1927), 360–366.

Bunge, M., *Causality*, Cambridge (Mass.), Harvard Univ. Press, 1959.

Cantor, M., *Vorlesungen über Geschichte der Mathematik*, 4 vol., 2nd ed., Leipzig, Teubner, 1901.

Crombie, A. C., 'Quantification in Medieval Physics', *Isis* 52 (1961), 143–160.

Euler, L., 'Introductio in analysin infinitorum', *Opera mathematica*, Vol. VIII–IX, Leipzig, Teubner, 1913, 1922.

Frey, L., 'Langages logiques et processus intellectuels', *Les modèles et la formalisation du comportement*, Paris, C.N.R.S. (1967), 327–340.

Grant, E., 'Nicole Oresme and his De proportionibus proportionum', *Isis* 51 (1960), 293–314.

Guilbaud, G.-Th., 'Des trucs et des machins: comment faut-il enseigner les rudiments de la théorie dite des ensembles?' *C. r. du Séminaire sur les modèles mathématiques dans les sciences sociales*, I (1961–1962), 20–28.

Hilbert, D. and Bernays, P., *Grundlagen der Mathematik, Vol. I, Berlin, J. Springer 1934*.

Inhelder, B. and Piaget, J., *The Growth of Logical Thinking from Childhood to Adolescence*, New York, Basic Books, 1958.

Inhelder, B. and Piaget, J., *The Early Growth of Logic in the Child*, London, Routledge & Kegan Paul, 1964.

Mac Lane, S., *Homology*, Berlin, Springer-Verlag, 1963.

Menger, K., 'An Axiomatic Theory of Functions and Fluents', *The Axiomatic Method*, Amsterdam, North-Holland Publ. Co., 1959, 454–473.

Milhaud, G., *Etudes sur la pensée scientifique chez les Grecs et les Modernes*, Paris, Alcan, 1906.

Mittelstaedt, P., *Philosophische Probleme der modernen Physik*, Mannheim, Bibliographisches Institut, 1963.

Newton, I., *La méthode des fluxions et des suites infinies*, Paris, 1740.

Papert, S., 'Sur le réductionnisme logique', *Etudes d'épistémologie génétique*, XI, Paris, Presses Universitaires de France, 1960, 97–116.

Peirce, S. C., *Collected Papers*, Vol. III, Cambridge (Mass.), 1933.
Piaget, J., *Traité de logique*, Paris, A. Colin, 1949.
Piaget, J., *Essai sur les transformations des opérations logiques*, Paris, Presses Universitaires de France, 1952.
Piaget, J., 'Problémes du temps et de la fonction', *Etudes d'épistémologie génétique*, XX, Paris, Presses Universitaires de France, 1966, 1–66.
Poincaré, H., *Science et méthode*, Paris, Flammarion, 1920.

CHAPTER 15

GENERAL CONCLUSIONS[1]

From these as yet very fragmentary facts, it seems that we may draw a number of conclusions which may not serve for the purposes of establishing an epistemological theory of functions, but rather as an 'Introduction' to such a theory. To go further we should need to pursue these studies within the framework of a more generalized study of the forms of explanation and of causality as is currently being undertaken in our Center. In fact, functions appear more and more to be the common source of operations and of causality. But, while the development of operations is becoming clearer, that of causality is still full of mystery. However, it does appear to be an attribution of actions and then of operations to objects. At any rate, a certain parallelism (although perhaps only a partial one whose significance is difficult to specify at this point) will undoubtedly end up by imposing itself between them. It is therefore only in this context of exchanges between operatory and causal structures that the notion of function will find a stable epistemological status. In the meantime, however, we can draw some conclusions from the preceding experiments.

1. THE ORIGINS OF FUNCTIONS

A function essentially expresses a *dependence*, whether it occurs between properties of objects which are variable or constant, or whether it is established between elements or characteristics which are inherent in actions or constructions of the subject. Even when defining functions as applications we do not disregard the idea of dependence, since the pairs involved are ordered, since the correspondence is univocal to the right, and above all since, as Grize says (§2), the true problem is that of the formative process which engenders the correspondence: in this case, the application of E onto F expresses a dependence of the target-set F in relation to its source-set E, otherwise the correspondence would amount to no more than a comparison and the application would be no more than the establishment of a relation, but not of a function. This dependence can only translate the operatory construction of the structure of F from E which is not always an ongoing or pre-existing dependence outside of the actions of the subject.

I. – This general characteristic of dependence, which distinguishes functions from simple relations (the latter result only from comparisons) thus raises the question of the origins of functions. Is their origin physical or operatory, or better yet, are they derived from the actions of the subject? If this is the case, according to which dominances, causal or operatory?

From the preceding facts, we can clearly see that each of these sources is found in turn for the functions studied. In the case of the jars in Chapter 7, the increasing or decreasing intervals are observed without having been predicted (stages I and II) which shows that the function is drawn from the observation of facts and that this occurs (at stage I) prior to any causal comprehension. Thus, in this example, the function is reduced to a simple physical law. In the case of the experiment involving pulling (Chapter 4), the lengthening of the spring as a function of the weight is also discovered by means of a physical experiment, but is immediately understood as a causal link. The functions involved in the case of the wheels and the cars (Chapter 11) are both physical and spatial, thus physico-geometrical, in the sense of the object space, while those which intervene in the transformations of a square into a rectangle (Chapter 8) derive from spatial operations which the subject could effect without the use of anything physical. The numerical differences involved in Chapters 5 and 6 are expressed as functions which are properly operatory, but are understood as relations which are always true even before the comprehension of the underlying reasons attains a deductive necessity which goes beyond this simple inductive level. Furthermore, before the operatory level is attained, the regularities of Chapter 3, §1, lead to functions resulting from a single action and its coordinators, i.e. the preoperatory action constitutes the source of subsequent operations. An action can also produce functions which are causal but not operatory as in Chapter 4, when the subject lengthens the spring by pulling on the string, thus bypassing the weights (for the question involving the mouse).

All the types of dependences can become functional, and when they do, the problem is then no longer to assign a specific origin for functions but to find what these multiple origins have in common. In this respect the status of functions is comparable to that of space, since spatial relations can also be drawn from objects by abstraction, whether it be simple or physical, or constructed through reflective abstraction on the basis of the general coordinations of actions founded on deductive operations when the latter use neighborhoods, bicontinuous correspondences, etc. There is nothing surprising about this kinship since all elementary physical functions comprise a spatial dimension; however, this does not solve the problem of isolating the characteristics common to the multiple sources from which functions derive.

II. – Before searching for this common source, let us dwell for a moment upon the two modes of abstraction from which functions derive. Logico-mathematical links are due to reflective abstractions because they are drawn from coordinations of actions and not of objects: even if the contents (in intension) of a class or a relation are drawn from objects (simple abstraction), their form (union, inclusion, order, etc.) is the result of an activity of the subject. Physical data, on the other hand, are drawn from objects. Where space is concerned we find a mixture of both: physical space is occupied by bodies and is also constructed operatorily by the subject. In this respect a function is analogous to space which is why like space it plays the fundamental role of mediator between deductive operations and causality, i.e. functions are the instrument by which physical laws as well as a large number of logico-algebraic structures (morphisms, etc.) are established.

A relation such as $A < B$, even when applied to two stones has no physical existence as such: the objects certainly exist with their lengths but they are not related until they are compared by the subject. The relation is thus physically very distinct from its terms, even if they are perceptually (illusion of contrast) or conceptually indissociable, since they are only indissociable for the subject to the degree to which their comparison is inevitable for him. By contrast, in a physical functional covariation (where A pushes B thus moving B, or in Chapter 4 where a weight is placed and stretches the spring, or in Chapter 11 where the same distance is travelled in the same time, i.e., $d = f(t)$, the link between the terms is a real dependence and not only a comparison. The terms are not independent of the function since their variations constitute a covariation, not only for the subject but also objectively.

As for functions which are logico-mathematical and no longer physical, the same dependence and coproperty or covariations still exist as a result (in contrast to the preceding ones) of the variations due to the operatory manipulation by the subject. In chapter V, the number of elements in set E_2 increases as a function of the decrease in the elements of set E_1 and we can therefore speak of a logico-mathematical function because these very simple increases and decreases (even if they are effected physically with tokens used as concrete symbols) are due to the operations of the subject. But if these two sets consisted of balls placed on the two scales of a balance with a device making it possible to pass balls from one tray to another, the covariations and dependences would once again constitute a physical function while the link between the subtractions and additions of chapter VI still represents a logico-mathematical function.

III. – Returning to the common source of these two types of functions, a fundamental fact stands out from the preceding experiments, i.e. the constitution of functions precedes by far that of operations if we define the latter in terms of their reversible compositions and in terms of the constitution of structured wholes which simultaneously imply invariants and a closed system of transformations. This is why an application which establishes a correspondence is such a precocious function even though correspondences don't become operatory until about age 7. Thus there exist preoperatory correspondences (without conservation of equivalence, etc.) which are already functions, and there exist functional compositions (cf. the ordered pairs of Chapter 1) well before the constitution of operatory structures, even of the most elementary ones such as groupings. Of course this classification is somewhat arbitrary since we can speak of 'prefunctions' as preoperatory behaviors defined, among other criteria, by their essentially qualitative nature, and we could then reserve the term function for quantified functions which result from operatory constructions.

On the other hand there are two reasons which suggest why we should consider preoperatory functions as *constitutive functions*[2] and quantified functions as constituted functions. Firstly, functions go beyond the domain of the subject's operations since they are in reality the common source of operations and of causality. There is therefore an advantage to considering functional 'dependences' in and of themselves before they are linked to operations. But the second reason is even more important. Psychologically, the common source of operations and of causality is constituted by the actions of the self whose dynamic aspects enable the subject to experience, through simple abstractions, the first links to become causal, while the structures of their coordination give way to reflective abstractions thanks to which operations are constructed. Now, an action comprises a set of ordered dependences (between the conditions under which it is effected and its results; between the objects serving as means and the terminal objects as in instrumental behaviors; between one action and its successor in more or less special or general coordinations, etc.). It is thus not only natural, but indeed helpful, to consider constitutive functions as the expression of the dependences proper to schemes of actions, with the assimilation of objects to these schemes representing a specific type of 'application.'

Given the above, we can understand why elementary functions precede operations and extend well beyond their domain and then subsequently encompass them. The first reason for this overextension is that not all actions are interiorized into operations since they are not always reversible nor always composable, it only being possible to interiorize general coordinations of actions (union and inclusion, ordering, etc.). Built into the composition of functions (Mac Lane's categories, Brandt's groupoids, Menger's hypergroups) are limitations (operations not everywhere defined, restricted transitivity, etc.), which goes without saying, if functions do in fact translate the schemes of actions and not their general coordinations. On the other hand, while some 'prerelationships' on the preoperatory level are already functions, only the compositions of relations, in the strict sense of the term, are operatory.

Above all, operations presuppose a differentiation between extension and intension and a quantification of extension (see the end of §4). Even in a given relation aRb the terms a and b constitute the extensional field of the relation which is in turn characterized by intension. Constitutive functions on the other hand are essentially qualitative or ordinal. See, for example, the description of the first stage reactions to each of our tasks, where the only attempts at quantification, when actually made, are either equalities by correspondence or ordinal additions when a proportionality imposes itself. Given a functional pair (x, y), we cannot say, as in the case of the relation aRb, that the objects constitute the extension and the function the intension, since the function does not link the objects as such[3] but only their properties. While it is true that these properties are qualified, which can be ascribed to intension, they are also attached to objects, which derives from extension, without the objects as such being terms of the function (they are terms of the 'category' but a category is not a system of pure functions since it constitutes a set of objects *with* their functions).

It is true that constitutive functions are numerically or metrically quantified (proportionality, etc.) which transforms them into constituted functions. However, constitutive functions as yet only correspond to globally conceived actions, while constituted functions can comprise quantifications of multiple forms (cf. all of analysis), but this higher stage presupposes a close cooperation (even to the point of identification) between functions and operations.

2. SCHEMES OF ACTION, COORDINATORS AND FUNCTIONS

If the source of constitutive functions is to be found in schemes of actions with regard to both its causal destinations and its operatory evolution, it must be possible to analyze the mechanism by which functions are formed from this viewpoint. This is what we intend to do based on the preceding facts, from the standpoint of the operatory rather than the causal aspects of functions.

I. – We call a *scheme* of an action that which makes it repeatable, transposable or generalizable, in other words, its structure or *form* as opposed to the objects which serve as its variable contents. But except in the case of hereditary behaviors (global spontaneous movements, reflexes or instincts), this form is not constituted prior to its content. It is developed through interaction with the objects to which the action being formed applies. This is truly a case of interaction for these objects are no longer simply associated among themselves through an action, but are integrated into a structure developed through it, at the same time that the structure being developed is accommodated to the objects. This dynamic process comprises two indissociable poles: the *assimilation* of the objects into the scheme, thus the integration of the former and the construction of the latter (this integration and this construction forming a whole), and the *accommodation* to each particular situation.

Assimilation, which thus constitutes the formatory mechanism of schemes (in a very general biological sense, since organisms assimilate the environment to their structure or form, which can in turn vary by accommodating to the environment) appears in three forms. We will speak of functional assimilation (in the biological sense) or 'reproductory' assimilation to designate the process of simple repetition of an action, thus the exercise which consolidates the scheme. Secondly, the assimilation of objects to the scheme presupposes their discrimination, i.e. a 'recognitory' assimilation which at the time of the application of the scheme to the objects makes it possible to distinguish and identify them. Lastly, there is 'generalizing' assimilation which permits the extension of this application of the scheme to new situations or to new objects which are judged equivalent to the preceding ones from this standpoint. Let us further recall that the coordination of schemes necessitates a reciprocal assimilation. For example, the coordination of vision and of prehension at the sensori-motor level at 4–5 months presupposes that the objects assimilated until then as 'to be looked at' are from then on assimilated as 'to be grasped'; the reciprocal can also be verified (in general immediately) in that an object outside of the visual field is now brought before the eyes and no longer, as was the case before, to the mouth.

Having reviewed these psychological notions, let us return to constitutive functions and attempt to relate them to the schemes and above all to the assimilation of objects to schemes, since it is assimilation which determines the formatory process. We can in this respect distinguish three questions or three successive planes of analysis: (1) Given any scheme of action (however varied the schemes and their contents), we must first characterize the various modes of functioning of this scheme which in turn leads us to distinguish the different *coordinators* at play in the functioning (this being the term we use for the combinators which link the successive actions deriving from the same scheme); (2) Since, on the other hand, the assimilation of objects to a scheme consists of integrating them into an organized whole, we must next examine the form of this *application* of the scheme onto the objects, which leads us to characterize the links introduced or observed between the objects: these links then constitute *functions* which, it should be noted, are introduced or observed (or both), depending on the diverse equilibriums in play between assimilation and accommodation since the links subsumed by the scheme can be as easily discovered among the objects assimilated to the scheme as added by the assimilation itself; (3) Finally, there remains the question of the coordination of the schemes among themselves (and no longer only of successive actions within a given scheme). This is expressed by the problem of the *composition* of functions.

II. – As regards the functioning of the schemes, the three forms of the formatory assimilation of these schemes distinguished in I are three of the coordinators corresponding to the principal operators distinguished by the combinatory logic of Curry, Feys, etc. which are used in Chapter 3 (§ 1) in the analysis of the regularities of certain schemes of action.

In effect, the coordinator *W* or *repeater* corresponds as naturally to reproductory assimilation as the coordinator *I* or *identifier* does to recognitive assimilation. As for the *permutator C*, we have seen that if we wish to express the elementary and general forms of the schematism of the action, we should distinguish the degrees or hierarchical types of permutation (as there already exist different degrees of identification). Given two objects A_1 and A_2, we will say that a simple substitution C_0 occurs if A_2 is chosen in place of A_1: this is the coordinator which intervenes in chapter I when one flower or train is substituted for another which is then disregarded. We will speak on the other hand of a C_1 permutation if A_2 is substituted for A_1 and vice versa. Similarly, the 'inversal' C_2 will refer to the substitution of A_2A_1 for A_1A_2, etc. Thus, it is clear that the simple substitution C_0 is the combination which intervenes in all 'generalizing assimilations', the generalization of the scheme consisting in the application of the scheme to the new substituted objects.

As for the associative coordinator B, it necessarily intervenes as soon as the scheme is applied to at least two objects at the same time, in which case the associator B produces ordered pairs, etc. In our analysis of assimilation, we did not deem it necessary to distinguish a special form for this case, however it goes without saying that if we want to make everything explicit, as is the logician's duty, we must add this combinator to the preceding ones: this is simply due to the general conditions governing the construction of schemes through assimilatory integration.

III. – A scheme of assimilation, in the course of its construction as well as in the transpositions effected by it in the repetition of an action from one situation to another, leads to *applications* in two concomitant and indissociable ways. There is, on the one hand, from the time of its construction, the application of some type of action (for example finding, displacing, modifying) to at least one object; and there is, on the other hand, a creation of links between two or more objects, simultaneously or by successive transpositions, which once again constitute applications. In fact, linking two objects as being assimilated to the same scheme is the same as applying that same scheme to them and establishing a correspondence between the characteristics of one of the objects (relative to the assimilatory scheme as well as to the object itself) and those of the other (according to the same interdependence): i.e. this correspondence is an application.

In short, the application of a scheme to objects, in the course of the assimilation which forms (in the sense of integration) the scheme as well as during the transpositions which also consist in the assimilation of objects among themselves, comes to be the same as constructing or utilizing (and in both cases observing) the dependences. It is these dependences which constitute *functions*. Whether we define a function as an 'application', as an ordered pair, or as a univocal correspondence to the right, it intervenes from the outset in the constitution of schemes of assimilation by organizing the actions and incorporating the objects. As a constitutive function it expresses this organization or better yet the links that it establishes.

IV. – There remains for us to consider the composition of different schemes through the reciprocal assimilation of these schemes. The process of this coordination by reciprocal assimilation sufficiently demonstrates the continuity that exists between the internal constitution of the scheme and its composition with others. From the standpoint of the composition of functions such a fact is quite significant for if we use the term 'category' to denote a set of objects together with their functions and we characterize its laws of composition by specifying that the operative link is not always defined, that when it is defined it is associative and that it nonetheless comprises an order (since identifications end up by distinguishing an undifferentiated element to the left and to the right), we find no continuity (as Grize remarks) between the original functional ordered pairs and the categories formed by several ordered pairs. This was confirmed as early as Chapter 1 with respect to the composition of ordered pairs.

In other words, functions are composed well before strictly operatory structures are constructed. This can be psychologically explained if functions express the links proper to schemes of assimilation of actions and if the coordination of these schemes is still a product of assimilations in the generalized form of reciprocal assimilations. Furthermore, these compositions only translate incomplete or partial coordinations while operatory structures are drawn by reflective abstraction from the most generalized coordinations of actions. From this we can infer the probability of a continuous but slow progression in successive multiple stages between the composition of functions and the construction of operations (in the sense in which we use this term). We focused on precisely these levels of gradual development in several of the preceding chapters, e.g., in Chapter 2, on the constitution of hierarchically included classes starting from the initial ordered pairs; in Chapter 3 and Chapters 8 to 12, on the construction of proportionality; in Chapters 5 and 6, on the composition of numerical addition, subtraction, duplication and halving, etc. In each of these cases, one of the fundamental characteristics of the passage from functions to operations or (consequently) from constitutive functions to constituted functions is the subordination to a progressive quantification (of the extension of classes, or of differences which become proportional, etc.). This occurs at the elementary levels where functions still remain essentially qualitative or proceed by equalizations due to the applications by correspondence. But prior to examining the process of quantification, we should analyze a second fundamental aspect of constitutive functions, i.e. their causal aspect.

3. FUNCTIONS AND CAUSALITY

If functions do in fact constitute the common source of operations and of causality we must expect that starting from the causal aspect of actions, functions which are isomorphic to the preceding ones will evolve along parallel lines, but based on the dependences existing between objects independently of ourselves, instead of on those which are engendered by our manipulations and operations.

I. – First of all, we must briefly explain what we mean by the term causality (which has as many meanings as the term operation) and why it proceeds from the actions of the self as do operations, even though it bears on the relations between objects and not on the constructions of the subject.

All actions modify objects. Even when the only purpose is to find an object, one must push aside obstacles, displace bodies (including one's own) thus transforming reality in some way. Consequently, when actions lead to operations they do so by abstracting out from this causal aspect so as to retain (1) only the structural and not the dynamic organization of the schemes of assimilation from which actions derive, and above all (2) only their general coordinations. On the contrary, if as a first approximation we define causality as a system of objective transformations comprising invariants, the actions of the self quickly provide us with models of causality, particularly in the case of instrumental actions: the child understands at an early age that the movements of his body are transmitted to an instrument (stick, etc.) and from there to external objects by means of certain transformations (movements, etc.) and certain limited conservations (the transmission of the pushes is compensated by resistances, etc.). Now, observation shows that (starting from the elementary sensori-motor levels) it is precisely these models drawn from the actions of the self which permit the comprehension of the causal relations between objects and also serve as the schemes of assimilation for the construction of reality.

But the dynamics of objects are not all causal and the plane of laws must be carefully distinguished from that of causes. Laws are drawn from repeated observations of objects and the covariations which they express can be discovered well before they are causally understood (whether the search for laws does or does not presuppose the search for causes remains an open question from the heuristic point of view). Once some laws have been established (and this occurs from the sensori-motor level on: a baby very quickly knows that a sound corresponds to a figure, that a suspended object can be balanced, etc.) they still need to be explained, and it is at this point that causes intervene: not as discovered (or discoverable) isolated laws, but as a system constituted by the coordination of laws. Such systems arise precociously from the actions of the self which are also necessary for the establishment of laws (in the preceding examples: turning the head to listen, touching the object to balance it, etc.). As a result causality is constantly being oriented or structured through actions in two stages.

The first stage is preoperatory. Here causality appears as a system of actions attributed to objects, in response to the actions of the self exercised on the object and corresponding to them. That is why when the child knows how to move a ball B by throwing a ball A against it in an instrumental action, he will immediately attribute to A the power to displace B even if A is not controlled by his own body (this is what Hume forgot in his famous billiards example when he neglected to take into account the cue as well as the player himself . . .). These actions, attributed to objects, can at the preoperatory level be of any type (mechanical, finalistic, animistic, etc.), and must be carefully distinguished from the actions of the subject which are necessary for his empirical investigations and for the establishment of laws. These latter actions are in no way attributed to objects but are only used in their exploration.

At the operatory level where causality is refined and made more exact, the distinction becomes even clearer. On the one hand, once operations have been constituted (classifications, seriations, serial correspondences or correspondences between classes, etc.) they accelerate the discovery and the establishment of laws by acting as operations *applied* to objects or to representations or symbols, etc. On the other hand, causality, as a system which explains these laws, gives rise to operations *attributed* to objects, which are themselves considered as the operators which transform situations while preserving the invariants. For example, the child explains the conservation of the matter, weight and volume of dissolved sugar through the hypothesis of small invisible grains which is based on additive operations, etc.

II. – The problem is now to establish the relations between constitutive functions and the set of instruments of physical knowledge formed by laws and causes. We will refer to this set from now on as the causal aspect of functions, as opposed to their operatory aspect, which is logico-mathematical in nature.

The first point to be made in this respect parallels exactly what we saw in § 2-II, i.e. in terms of their physical or causal significance, schemes of action evidence certain general modes of functioning which can be characterized as *coordinators* and which correspond term-by-term with the combinators of combinatory logic. However, in this case they are attributed to objects instead of being used only in the thought of the subject:

(1) The coordinator W' or *repeater*[4] naturally intervenes in the establishment of laws but already with a sort of causal aura since for the subject it is the phenomena or the behaviors of the object which are expected to repeat themselves. This is why, in Chapter 4, the youngest subjects are convinced at the beginning that if we replace a weight, the spring will stretch as previously observed, etc.[5]

(2) The *identifier* I' is also attributed to the objects as such, which are conceived as conserving their identity at the same time as they are identified by the subject. Likewise in Chapter 4, when the youngest subjects are presented with a string in two segments y and y' (at a right angle) of which one segment (y') becomes longer when the other (y) becomes shorter, they

admit that it is 'the same string' which goes from y to y' well before they acknowledge the conservation of its length.

(3) The *permutator C'* has a very general physical significance as a displacement combinator. At the ordinal (preoperatory) level, movement is not conceived in metric terms (the interval covered), but simply as a change in position: a movable unit A is thus considered, in relation to a non-movable reference element E, as passing from the order AE to the order EA. It is this permutation of positions which characterizes displacement before hyperordinal and metric considerations bring into play the relative size of the intervals and then their quantification (see §5 below).

(4) The *associative* coordinator B', expresses the general action of union, but once again the action is attributed to objects and this time intervenes well before operatory compositions and their conservation. In Chapter 7, for example, 5–6 year-olds observing the different parts or levels of water which drain successively from jar A into jars B and C, do not doubt that the union of these associated quantities would again result in the same amount (they even expect a total isomorphism of the levels and intervals, etc., without subsequently understanding what happens due to their lack of operatory composition and conservation). In this case, the associator B' effectively signifies a union which occurs well before operations.

III. – The observations, anticipation and generalizations of repeatable relations between objects give rise to *functions* of which Chapters 4, 7, and 11–12 provide numerous examples. In fact, from the outset these links have the fundamental characteristic of expressing real *dependences*, i.e. ordered connections such that when $y = f'(x)$, y is subordinated to x (cf. the dependence of the spring's lengthening on the weight in Chapter 4, the subordination of the filling of jar B to the draining of A in Chapter 7, etc.). These functional dependences, insofar as they are physical, constitute what we can call laws,[6] however imprecise and relative to the subject's level of development they may be, since we are dealing with objectified functions which express the coproperties or covariations of objects and not simply those of the terms of the subject's mental manipulation.

The problem of discovering the relations between these functional laws and causality is very general in nature. But before posing and discussing it fully, we must emphasize the following, i.e. the physical combinators W', etc. of which we have just spoken (in II) and the objectified functions or laws which they make possible, do not replace but only duplicate the general combinators W, etc. of §2 in II, which we shall call 'cognitive coordinators' and the functions which derive from these (§2 – III) which shall be called 'cognitive functions'. In effect, to establish a law, i.e. to acknowledge that nature repeats itself (W'), the subject must first repeat his own exploratory actions and observations (W) and it is in 'applying' the repeater W to reality that he will subsequently be able to 'attribute' it to objects or phenomena which are themselves of the form W'. In order to recognize that the objects under consideration remain identical (I') he must first of all identify them (I). Likewise, to judge a displacement C' of the object he must permutate his successive observations, and thus must use C; and in order to consider elements as united (B') he must begin by associating them by B. The result is that when the subject acknowledges a physical or objective dependence $y = f'(x)$, he must begin by using and 'applying' a mental or cognitive dependence $y = f(x)$. The significance of this dependence is that to know or determine y he must first know x.

It can be said (for the most part with reason) that a law is not explanatory in and of itself since it is limited to establishing a regular succession of observables, and this holds true for $y = f(x)$. But there remain two questions to be solved for the significance of the actual law $y = f'(x)$ to be understood insofar as it is physical: (a) What are the new properties added by the physical coordinators W', I', C' and B' to the cognitive coordinators (W through B); and (b) given that the real or physical dependence is acknowledged very precociously by all of our subjects, what is the basic significance of the dependence between y and x which is judged to be real in $y = f'(x)$ with respect to the *notional* dependence in $y = f(x)$?

As regards the physical coordinators W', etc., it is remarkable that all are 'attributed' to objects, and if our hypotheses are correct, this already marks a step in the direction of causality. The coordinator W' is already significant in this respect since, for the subject, it is reality which repeats itself and not just perception or language. Now, this permanence of laws further implies that objects conserve their identity (I'), that they are displaced (C') and that they are united (B'), independently of the subject. Thus in contrast to the set of coordinators which simply express the activities of the subject, the set of physical coordinators manifests an ontological or ontic point of view which goes beyond phenomenalism and postulates the idea that there is 'something' beyond the observable whose presence must be taken into account. And although this certainly does not comprise a causal explanation since the 'how' is lacking, the hypothesis does make necessary a causal explanation.

The idea of dependence leads to the same conclusions. There exist at least two if not three types of dependences: (1) a 'notional' dependence such that in $y = f(x)$ the knowledge of y depends on that of x; (2) a 'physical' dependence such that, for example, the movement y depends on the shock x made by an external object; and (3) a dependence relative to an effective action of the subject (x being, for example, a thrust resulting from a manipulation). It goes without saying that type (3) is the common source of the other two and that sooner or later it is distributed in two directions from which point only forms (1) and (2) remain. And as we saw they intervene jointly and in a parallel fashion. What the notion of a real or physical dependence now adds is the idea that this 'something' situated beyond the observable by the combinators is not only present, but is also active and a source of connections: this brings us even closer to the actions 'attributed' to objects, thus to the exigency of causality. That is why, starting with the preoperatory level, an *ad hoc* causal explanation, traced or copied from the causality of the action itself is superimposed on these functional dependences, e.g., the weight 'pulls', it is 'strong', etc. In these cases, which have many analogies in the history of science (the force of attraction at a distance, etc.), we can echo Grize's remark that causality only belongs to the metalanguage and not yet to the language of functions. On the other hand, and in apparent verification of the generality of this hypothesis, the subject is also satisfied in certain cases with isolated functional laws (stage I of Chapter 4, etc.) established by pure observation and without causal comprehension.

IV. – How then is this causal comprehension constituted, from particular functional laws or from these isolated functions (isolated in the sense that they can be independently discovered). By their composition, of course, since causal explanations derive from a system of composed functions and not from a number of isolated functions. But the interesting point about this composition, an example of which was provided in Chapter 4, is that it simultaneously presupposes a cognitive or operatory organization of functions as well as the attribution to objects of a set of operators which provide the framework for the functional 'dependences' at the same time that they coordinate them. It is at this point that causality is incorporated into the language of functions because then it consists, not of a special function among others (the so-called 'causal laws' which do not exist as such) nor in a semantic and metasyntactic duplication of this or that function, but in an overall structure which is explanatory insofar as it is a structure to the degree to which it unites into a whole the transformations and conservations which alone can explain the *production* of observable phenomena.

The passage from laws, which with their coordinators and their 'dependences' already imply a universe of active 'objects' beyond the observable phenomena, to the causal 'production' of these observable phenomena, is dual in nature, or is at least pursued along two parallel paths like two armies which join forces to converge on a single objective. One of these paths is that of the cognitive coordinators $W \ldots B$, followed by the cognitive functions $y = f(x)$ and finally by the operations themselves, which permit the composition of the transformations and assure their conservations. Throughout Chapter 4, functional pairs are coordinated until they constitute a transitive series and the operatory composition of the relations drawn from these functions assures, through the interplay of compensation, the conservation of $(x + y)$ and of $(y + y')$ as well as the metric equality of the differences Δx, Δy, $\Delta y'$ and Δz. The other path, which is parallel to the first, consists in projecting the coordinators $W \ldots B$ onto the physical coordinators $W' \ldots B'$, the functions $y = f(x)$ onto the physical dependences $y = f'(x)$ and the operations themselves onto operators 'attributed' to objects; these thus permit the closure of the system of dependences into a structured whole which assures their production. In fact, the example in Chapter 4 involves the ascent of the forces from weight z up to the spring x and the descent of the displacements (lengthening of the spring, displacement of the string and lowering of the weights), with the closure and the circularity of the system accounting for the functional dependences observed.

But which of these two paths, operatory or causal, controls the other, orienting it from among other possibilities towards this parallelism? As much one as the other: operations can precede and give rise to causality even prior to any experimentation and observation of laws, as seems to have been the case with Greek atomism; but a hypothetical causal model can also impose itself on operations and compel them to imitate it in the same manner – to use a somewhat daring comparison – in which mathematical physics constructs its theorems, i.e. deductively, but in correspondence with the explanatory models inspired by experimentation.

The parallelism of these two paths applies as follows to the modest beginnings of the function studied in Chapter 4. The physical *dependences* discovered through experience, which in this case still only constitute functional laws and one-way functions, are subsequently coordinated as *interdependences* which then lead to a system of differentiated actions oriented in both directions, taking on by that very fact a causal significance. However, this elaboration is only made possible by the parallel construction of operations which encompass the initial oriented functions into a set of reversible transformations which lead to the notions of conservation and to a quantification of covariations (all the way to a structure of proportionality). When the interdependences finally reach a level of intelligible causality, we find a complete isomorphism between the pulling operators whose action is transmitted from bottom to top, or those of expansion and of displacement oriented in inverse directions, and the operations of the subject who deduces and constructs by implication these same links as elements of his interpretation.

4. APPLICATIONS, CLASSES AND RELATIONS

Having thus isolated (§§ 1–3) the *sui generis* nature and precocious importance of constitutive functions, we should now attempt to explain their passage to constituted functions, i.e. the progressive interaction of functions and operations.

In fact, constitutive functions represent the formatory matrix of future operatory structures, in other words, they express the way in which actions lead to operations. But even after the latter have been elaborated, the functions remain oriented 'dependences' or applications of structures and they are enriched by a progressive quantification which diversifies their varieties and above all permits them to be operatorily composed among themselves and to thus multiply themselves *ad infinitum*.

I. – The first point which must be understood involves the passage from functions to operatory class inclusions. Grize shows that from the logical point of view this link, exemplified in Chapter 2, is fairly direct. As in Chapter 2, let the elements $A'_{1,2,3}$ constitute a source-set which the subject will use to cover a part of base A (in this case the target-set). The application of A' onto A, constitutes a functional link (univocal to the right; by contrast, the correspondence of one A to several A'''s is not an application due to the lack of this univocalness) which defines a quotient-set on the set of the A'''s which in turn constitutes an equivalence class (since the A'''s are equivalent in terms of their result on A). On the other hand, if we extend the source-set E to include all of the A', B', C' and D' elements (1 to 3 on each) and the target-set F to include the four bases A, B, C, D, we note that the application of a quotient-set onto E only covers a subset of F, thereby introducing the notions of subclass and of hierarchical inclusion. It is thus easy to pass from functions to a system of included classes with their possible quantifications in extension and since these classes are disjoint, it is possible to compose them among themselves as a grouping of classes.

This passage occurs in three successive stages, the first of which (age 5–6) already introduces equivalences based on the 'suitability' of A' to A, etc., but only in the last (age 8–9) is the operatory system of hierarchical grouping attained. But now this logical passage must be translated into genetic terms so we can understand the relations of epistemologic filiation between applications or preoperatory functions and the operatory structures of equivalence classes. Psychologically, the slowness of this passage is due, as was shown in Chapter 2, to the fact that here again the initial equivalences only result from an assimilation of objects to the schemes of actions, i.e. in this particular case, to a scheme of application by covering (recall that this initial stage is characterized by definitions 'by usage'). It is only in the course of stage II that equivalences begin to be formed through an assimilation of objects among themselves, but as a function of the scheme: from this result the small 'individual unions' of elements (= the 'partitions' of mathematicians) without as yet any hierarchical inclusions. The latter finally appear in stage III (that of definitions by kind and specific difference) but are based on the reversible operation of the union of classes and no longer only on individual elements (thus on the operations $\alpha + \alpha' = \beta$ and $\alpha = \beta - \alpha'$).

But while this evolution confirms the primitive and preoperatory nature of applications and of functions as the expressions of schemes of assimilation of actions, it shows above all that the passage to the operations of inclusion of equivalence classes presupposes a new process, i.e. reversibility, whose intervention here is perfectly clear. Actions as such are one-way and thus are not operations (in the limited sense we normally use): the application of E onto F constitutes a correspondence of many to one (Figure 18, I) which is a function in that it is univocal to the right. On the other hand, the one to

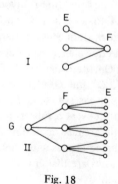

Fig. 18

many correspondence of F to E is, by definition, neither an application nor a function since it is not univocal to the right. Yet, it is precisely this one to many correspondence (for example, from G to F and from F to E in Figure 18, II) which characterizes a hierarchical classification (a genus and its species, these species and their subspecies, etc.), and, to attain it, the subject must be able to go from one direction to the other without difficulty in this system of co-univocal correspondences (one to many or many to one). It is therefore only in freeing himself from the one-way direction of functions that the subject attains the reversible mobility of operations.

II. – Recall that the complementary subclasses (in Chapter 2 the elements A', B', etc.) are not determined by the applications of E onto F. Now we understand why. In the class union operations $\alpha + \alpha' = \beta$ and $\alpha = \beta - \alpha'$, the complement α' is univocally determined because the operation is reversible while this condition which is necessary for the constitution and inclusion of complements is not met in the case of one-way applications.

It should also be noted that while the construction of functions is linked to the use of coordinators and then of combinators (as we said in §2 above, but without the 'application' itself being one of the coordinators, for the latter characterize the formation of the scheme while the former characterizes its utilization in the assimilation of objects), the essential difference distinguishing these coordinators from operations (more correctly called operators) is that the former remain irreversible as do actions in general, while the latter are reversible. We can of course attenuate this opposition on the logical plane (by introducing an inversion operator or by suitably manipulating the permutator C), but from the standpoint of genealogical filiations, however abstract, the conservation of this very revealing contrast is worthwhile.

On the other hand, if we can distinguish two modes of classification (on which Grize rightly insists), one by the direct union of individual elements (the 'partition' of mathematicians) and the other by the subdivision of the initial set into included classes and subclasses, we can clearly see that this distinction also applies to the instruments of construction: individual unions are related to one-way applications and subdivisions are related to reversible inclusion operations. Since applications are much more precocious genetically, we can distinguish two levels, one preoperatory and the other strictly operatory (resulting in the quantification of inclusions and in the univocal determination of complements). It is instructive to observe that in making spontaneous classifications, the child passes by himself and in a fairly general manner from the first mode to the second. The latter imposes itself by virtue of the progress of the operatory structures which incorporate the constitutive functions after having resulted from them and which go beyond them in generalizing the co-univocal correspondences in both directions. But, while surpassing them in most cases, they refine them and thus lead to constituted functions, thanks to various methods of quantification to which we will return later.

III. – There remains for us to clarify the relationships between functions and relational groupings. A function expresses a dependence while a relation is only the result of a comparison. The difference can be most clearly seen on the physical plane, where the dependence is objective and the comparison is due to the subject, while on the logico-mathematical plane, when comparison is oriented it approaches the univocal dependence to the right established by an application. But, the difference is still clear from the formal standpoint since a function comprises an ordered, thus one-way pair, while a relation either has no order or it establishes one between the terms but one which can be traversed in both directions.

We can take as an example the grouping of serial correspondences, such as the special case studied by N. Van den Bogaert in our center.[7] A truck passes in front of 5 houses following a complicated course, takes different colored tokens from the front of each house and places them in the back of the truck in the order that they were picked up. Two problems are then posed: (a) explain why the tokens have been placed in that particular order on the truck ('why is the red before the yellow?' etc.); and (b) state the order in which the truck passed in front of each house. 60% of the 5 year-olds (15 out of 25) and 68% of the 6 year-olds (17 out of 25) solve questions a, while only 28% of the 5 year-olds (7 out of 25) and 44% of the 6 year-olds (11 out of 25) understand problem b (at age 7–8 we find only 4 out of 50 subjects who solve a without b). It is thus clear that prior to age 7, the oriented function which makes it possible to answer question a is not accompanied by the reversible operation assuring its reciprocal in b. This is an excellent example of a serial correspondence which is not as yet reciprocal or of a bimorphism which is not as yet an isomorphism (a situation which is currently accepted genetically speaking but which only occurs in advanced mathematics, as Guilbaud noted when these facts were being discussed.

Thus it is once again reversibility which distinguishes functions from the operatory structures of relations but there are also other differences besides the inverse operation required by the relational groupings. Given a 'category' of ordered pairs, where arrows stand for relations, these arrows give rise to partial compositions (or 'operations' not everywhere defined) but not to a strict transitivity. On the other hand, this system does not comprise one general identity, but two, by elements, since even the pairs (a, a), etc. are ordered, which necessitates the distinction of left and right identities. The grouping of relations, in contrast, ignores this difference due to its reversiblity and thus includes the laws of reabsorption and tautology missing from 'categories' of pairs.

We can thus generally acknowledge that the basic *décalage* which separates constitutive functions from the development of elementary operatory structures of 'groupings' is due to the gradual formation of the reversiblity of class unions and of concatenations or transitive sequences of relations, these reversible operations of union or concatenation thus being the source of the inclusions which are missing in the initial 'categories' of ordered pairs. Now the reversible inclusion (since it can be inverted into de-inclusions) leads to a quantification of the increasing or decreasing extension of classes with their equivalence relations which are themselves more or less broad or narrow, and of the increasing or decreasing difference (or distance) among the terms $A, B, C, \ldots$ in a concatenation of asymmetrical relations. This quantification in turn distinguishes the first essentially qualitative constitutive functions from the class and relational groupings whose quantification necessarily imposes itself in a form which is at first intensive before their synthesis leads to numerical or metric quantifications.

IV. − Before analyzing this process of quantification, let us restate some definitions. An *intensive* quantity is one which characterizes only the relationships of a part to the whole and ignores the quantitative relations between the parts themselves. The intensive quantity is thus limited to expressing the fact that there are more elements in a non-empty class B than in one of its subclasses A or that in the seriation of three terms $A < B < C$ (or $0 < A < B$) the difference between A and C (or between 0 and B) is greater than that between A and B (or between 0 and A), this being so simply by virtue of the nature of inclusions. We will retain the terms 'intensive quantity' (which is due to Kant) and 'extension' when used to designate a set of individuals united in one class, in spite of the ambiguity (purely verbal) which may arise from the fact that 'extensional' relationships only bear on 'intensive' quantities (this ambiguity is compounded in English where *'comprehension'* is translated by 'intension'). On the other hand, an extensive quantity designates the quantitative relationships between the parts of a single whole (e.g. the 'almost all' of Kowalevska), or the relationships between distinct totalities, when they are only expressed in terms of more, less or equal: such as when we are able to judge by some method (correspondences, etc.) that in a class $B = A + A'$ the complement A' contains more, less or as many elements as the subclass A; or when, in a seriation $A < B < C$, we can evaluate the difference between B and C as being larger, smaller or the same size as the difference between A and B (Suppes termed this non-numerical estimation of the intervals 'hyperordinal'). Finally, we will use the terms *metric* or *numerical* quantity for any comparison based on an iteration of cardinal units. This form of evaluation is a subvariety of the 'extensive' type.

5. THE QUANTIFICATION OF FUNCTIONS
AND THE PASSAGE TO CONSTITUTED FUNCTIONS

We call 'constituted' functions the countless differentiated functions which are developed through their interaction with operations. Their most general characteristic stems from their passage from qualitative coproperties resulting from elementary 'applications', to operatorily quantifiable covariations, then to variations of variations, etc. Chapters 3, 5–6, and 8–9 provide us with numerous examples of this evolution, not to mention the parallel development of the physical functions examined in Chapters 4, 7 and 11–12.

I. – Such a development begins by leading from qualitative (or even additive in an ordinal sense) 'preproportionalities' to metric or numerical proportionalities with this process itself being subordinated to the constitution of hierarchical inclusions (additive and above all multiplicative).

A proportion is an equality of relationships, in other words, a specific case of equivalence between relations. Consequently, we can use the term 'preproportionality' (or even 'correlate' in the sense used by Spearman although we sometimes broaden this notion) to refer to equivalences of all types, including functions. The significance of such a concept, however broad it may be, is simply that it emphasizes the fact that proportion in the strict sense constitutes the point of arrival of a long sequence of compositions which begins with the composition of two ordered pairs. Let a and b be two objects assimilated to the same scheme by virtue of any application, whether because their respective properties α and β resemble each other or because these objects 'suit' each other for whatever reason; we thus write $\beta = f_1 (\alpha)$. Now suppose that the objects c and d are linked in an analogous pair by virtue of an application f_2 which is more or less equivalent to the first ('more' would imply the identity $f_1 = f_2$ and 'less' involves the fact that both functions are coordinated by the same subject). Thus in any case, if γ and δ are the characteristic properties of c and d, we have the preproportion 'δ is to γ as β is to α'. This is not as trivial as it might seem, for this link marks the beginning of the most elementary of compositions: the coordination of two successive actions or of two functional pairs, whether similar or different. In Chapter 1, for example, when the child substitutes a small blue flower for a small red one, then a large blue one for the small one, he is already coordinating two substitutions, and he understands that in spite of their different contents, these two exchanges are comparable and therefore permit the substitution of the third flower for the first, thus allowing him to obtain a large blue flower for a small red one (naturally this sort of transitivity is not always possible).[8]

Starting from these entirely general and elementary preproportionalities, the subject arrives at preproportions which are differentiated and better defined as the equivalences between f_1 and f_2 are made more precise. Grize gives several such examples of which the most typical are the 'correlates' resulting from a multiplicative matrix of classes, i.e. by the Cartesian product of the classes of fruit with those of the vegetable families: 'grains are to wheat as grapes are to the vine', etc.

II. – Such a multiplicative structure rests on a set of one-to-one (biunivocal) correspondences. It is with regard to serial correspondences that we can most clearly see the conditions of the passage from preproportions to actual proportions. We note in passing that in thus using serial correspondences we do not leave the domain of inclusions, for in any given order $A, B, C, \ldots$ we can always consider the relation A, B as included in the relation A, C, etc., in the same way in which a subclass is included in a class of greater extension. Chapter 2 (§3), among others, gives us a good example of this passage.

In that experiment the children are asked to establish a correspondence between three fish of varying lengths such that $A < B < C$, and the corresponding amounts of food (for example biscuits of different lengths A', B' and C'). The stages observed in this task display the kind of progressive development under discussion. In what follows, however, we will not adhere to the values used for A, B and C in that experiment; rather we define a as the difference between A and B; b is the difference between B and C, etc.; and a' is the difference between A' and B'; b' is the difference between B' and C', etc. As long as $A < B < C$, a, b and c can be interpreted in any way.

(1) The first stage comprises the simple recognition of the existence of these differences (thus the discarding of the equalities $A = B$, etc.), and the arbitrary correspondence made between the differences a, b and c and the differences a', b' and c' ($b' > a'$ or $b' \leqslant a'$, etc.). There are elementary preproportions of the form '$B' > A'$ as $B > A$' in this first stage but without any attention being focused on the amount of the difference.

(2) The second stage is of the type $B' = A' + 1$ and $C' = B' + 1$ which is the same as saying, in general terms, that to the a, b, etc., which still remain subjectively undefined, there correspond the differences $a' = b' = 1$. We could thus speak of additive preproportions though in reality this addition is essentially ordinal: it comes to be the same as admitting that if A, B and C occupy ranks 1, 2 and 3, and the corresponding terms A', B', C' do so also,

then the differences a', b', etc. are each equivalent to 1 additional rank and are thus equal among themselves (but there is no cardinal quantification beyond $a = b = \ldots$).

(3) The following stage is on the other hand 'hyperordinal': $a' \leqslant b' \leqslant c'$ ..., or $a' = b' = c' = k \pm m$. The essential progress evident in this third stage is the attempt to take into account the value of the variation, and as soon as the child realizes the importance of this he finds himself on the threshold of proportionality. Up to this point, his preproportionality was reduced to: B' is to B as A' is to A, in the sense of a simple coordination of pairs without as yet any quantified differences. From that point on, on the contrary, the subject arrives at 'b' is to b as a' is to a,' considering the actual intervals between A and B, then B and C, seeking to carry over the value of the differences a, b, etc., onto a', b', etc. But these differences are not yet metrically determined and are still only evaluated in terms of 'more' or 'less', i.e. precisely in a hyperordinal manner (in the sense used by Suppes). Furthermore, the subject does not base himself on $a' = a$, $b' = b$, etc. but rather seeks to transpose a, b, etc. into a', b' in accordance with the value X' attributed to the original element A, B or C, and to adapt these differences to those which he observed on A, B and C. This marks the beginning of true proportionality, but is evaluated in terms of extensive and not yet metric quantities.

(4) Lastly, these qualitative proportions are made precise in a numerical or metric relationship: $a' = na$, $b' = nb$, etc., resulting in $b'/b = a'/a$ and the equalization of the crossed products is then possible.

III. – In their general form, and in particular when there is an inverse proportion between the differences a, b, c .... and a', b', c' (as in the case of the scales where an increasing distance from the axis is balanced by decreasing weights), metric proportions are not attained until about age 12 because they presuppose a group of four transformations: identity I, inverse N, reciprocal R and correlative $C (=NR)$, where the reciprocal can lead to the same result as the inverse N but by compensation and not by cancellation. By contrast, in a direct serial correspondence, when the proportion in a given increasing series with equal differencs must be reconstructed on another series with different values, the only transformations which intervene are those identical to themselves and their reciprocals without their inverses (neither N nor C).[9] This simplifies the problem and the solution is reached earlier. This is especially true in Chapter 2, §2, where the $1 : 2$ proportion of the holes can be obtained by additive compositions: to $1, 2, 3, 4 \ldots$ there correspond $2, 4, 6,$

8 ... by the repeated addition of 2. (This does involve multiplication but only implicitly since $1 + 1 = 2 \times 1$.) This is also the case in Chapter 2, §3, where the solution of the problem is more or less included in its own formulation since we tell the subject that B eats twice as much as A, and C three times as much. Interestingly enough, this does not keep the subjects from passing through all of the stages where the difference is first arbitrary, then equal to +1, then to $+k > 1$, before proportionality is attained.

The crux of the problem is the manner in which the subject succeeds in passing from qualitative preproportionalities to real or quantified proportions. The issue revolves around the emergence of hyperordinal proportions in which the subject comes to take into account the value of the intervals or differences by limiting himself once more to an extensive quantification which then rapidly becomes metric.

The answer seems to lie in the constitution of reversible inclusions which are the source of intervals prior to the emergence of a combinatorial system and the propositional operations which permit the attainment of the INRC group. At the preoperatory level (of arbitrary differences between simply ordered terms $A < B < C$ ... or differences equal to +1) the subject is limited to a very general structure of irreversible order which has no inclusions due to this very lack of reversibility. This situation is due to the primacy of constitutive functions (or is expressed by them) since all these functions are oriented (ordered pairs, applications, etc.) but are not operatory. Once class inclusions are attained due to additive operations, the included classes A in B, B in C, ... comprise (thanks to the reversibility $A = B - A'$) the complementary classes A', B', C' ... (see Chapter 3). Now the latter express precisely the difference or interval between A and B, between B and C, etc., which implies the intensive quantification 'if $A < B$ then $A' = B - A'$, etc. The same applies to ordered relations: if A is smaller than B and B is smaller than C, etc., then the relation $A < B$ is included in the relation $A < C$ and each of the relations $X < Y$ comprises a complement under $X < Z$ if $Y < Z$ is part of the same series. This does not tell us the size of the differences, although we can deduce, as in the case of included classes, that the difference $A < B$ is smaller than $A < C$ and that the difference between the two differences is equal to $B < C$.

This formation of complementary classes or relations thus constitutes an initial stage essential to quantification which comes to bring out the existence of intervals or differences between the including classes or relations and the included classes or relations. But the quantification involved is still 'intensive' (see the end of §4) and there remains the passage from there to the quantitative comparison of these differences or intervals which presupposes an 'extensive' quantification. Now it is a psychological fact that as soon as the intervals are noticed, they give rise to an evaluation in terms of 'more' or 'less' (i.e. hyperordinal and not yet metric). In accordance with the Estoup-Zipf-Mandelbrot Law[10] we know that this judgmental estimate is quite natural since it can be applied to both the classifications used in a store as well as to those used in biological taxonomy. The approximate proportionality indicated by this law between the size of the genus and the number of the species certainly comprises a comparison of the totalities among themselves and of the parts among themselves (thus an extensive quantification), based simply on an examination of the configurations of the whole. This is precisely what we observe in our subjects at a certain level when, given these configurations, they judge whether a seriation has equal or increasing differences, or whether the extension of the complementary classes is or is not larger than the primary classes, etc. These comparisons are 'applications' or varied correspondences.

It is therefore sufficient for these oriented applications to be completed in the form of two-way (thus operatory=reversible) correspondences so that these extensive proportions can become numerical and end up as strict proportionalities which will be generalized in the quaternality group.

IV. – However, proportionality is not the only characteristic of constituted or operatorily quantified functions. The most general example is that of the variation of a variation or of the functions of functions. Chapter 7, among others, showed how, after covariations succeeded coproperties and these covariations were understood, they led the subject to the notion of a variation of variations, which was first observed, then generalized, and from age 11-12, deductively anticipated. Two stages must be distinguished. In the first the variation of variations is not predicted although the observation of the facts leads to its comprehension and certain generalizations; this level which begins towards age 7–8 corresponds to the level of intensive and extensive quantifications analyzed previously. In the second stage these variations of variations are anticipated and deduced, i.e. finally completely assimilated: this second level corresponds to that of propositional operations which is natural since it is a question of operations on operations, i.e. of the second order operations which characterize the stage beginning at age 11–12; this results in the kinship between these structures and proportions.

The chapters written by Vinh Bang (8 to 13) provide us with excellent examples of variations of variations and of proportionalities. Chapter 8 is particularly significant. Given an L-shaped figure, the child understands, beginning at age 7, that the lengthening of one of the segments (x) compensates the decrease of the other (y) exactly and additively when the string sliding around a pin placed at the point of intersection between x and y is pulled. But if the figure is made into a square (with 4 pins at the 4 corners and this square is modified into a rectangle (by displacing the pins), it is not until age 12 that the subjects understand these simple additive compensations $(+ x = - y)$ and find the law illustrated by Figure 11. The fact that the square is a closed figure in contrast to the two single segments x and y, introduces the element of the surface area. For the function involved, which relates exclusively to the constant perimeter, to be discovered, the latter must be abstracted from the transformations of the whole which are also diminishing the surface area. This surface area (xy) is at a *maximum* when the figure is a square (i.e. x^2 since $x = y$) but then gradually decreases until it disappears when the rectangle is reduced to a line segment equal to half of the perimeter.[11] This decrease in the surface therefore constitutes a variation which corresponds as follows to the additive composition of the segments x and y (each being 10 cm at the outset): $10 \times 10 = 100$; $9 \times 11 = 99$; $8 \times 12 = 96$; $7 \times 13 = 91$ ...; $3 \times 17 = 51$; $2 \times 18 = 36$; $1 \times 19 = 19$ and $0 \times 20 = 0$. Thus it is due to his failure to dissociate the two compositions xy (surface) and $2(x + y)$ (perimeter) that the subject is not able to master the composition $2(x + y)$ until age 11–12 while he is able to do so at age 7 (but not before due to the absence of conservation of lengths and of exact compensation) for $x + y$ alone.

Chapter 9 provides another example of simple proportionality (circles to be ordered on sticks of variable length) where the solutions are not attained until age 11–12. In contrast, Chapters 10 and 12, given their causal context, show how the passage from simple dependences (the discovery of which sometimes occurs remarkably late as in the case where the distance travelled by a wheel is a function of its diameter) to interdependences (as between the distances, diameters and frequencies of the wheels) is accompanied by the operatory compositions needed for their constitution. These compositions begin towards age 8 through the development of quantitative conservations which duplicate the simple qualitative repetitions (the combinator W' referred to in §3) of quantified values. They then come to bear on the estimation of sizes of variations and above all on the comprehension of the relative sizes, with the quantification of the intervals or relationships between the differences leading, towards age 11–12, to proportionalities.

6. FINAL REMARK: FUNCTIONAL ORDER AND THE REVERSIBILITY OF INCLUSIONS – THE LOGIC OF COORDINATORS (ELEMENTARY COMBINATORS) AND OF OPERATIONS.

The two principal accomplishments of the preceding studies are that we were able to realize a dream shared by several of us, i.e. to isolate a logic (or a relatively coherent prelogic) of preoperatory structures; and to account for the unlimited production of 'constituted functions' in contrast to the limited number of operations.

I. – Constitutive functions express the links inherent in schemes of actions and are thus organized according to the common forms of these actions which we term elementary coordinators (W, I, C, B), instead of combinators, to emphasize this elementary aspect. Actions are oriented and often irreversible from which stems the ordered or oriented character of functions and applications.

This fundamental property of being ordered (univocalness to the right) of actions and functions thus explains (or expresses, depending on one's point of view) one of the most general traits of preoperatory thought (age 4 to 7): the primacy of order relations over inclusions in the most varied domains. This is why lengths are evaluated by the points of arrival (longer = farther) well before they are evaluated in terms of intervals: thus the importance of boundaries in all imaged representations; speeds are evaluated in terms of overtaking; classes begin as ordered (ranked) figural collections; numbers are also subordinated to the length of the ranks, evaluated in accordance with the over-extensions; reciprocities are poorly understood due to the predominance of one-way orientations, etc.

This primacy of order, which is naturally a source of all types of illusions (the non-conservations due to ordinal estimates, etc.; the non-equality of a path A, B, and its return path B, A, etc.), comprises a valid logic where justified: the logic of constitutive functions and of elementary coordinators or combinators (repeaters, identifiers, permutators in the form of substitutions, etc.). This logic presents the distinctive characteristics of all logic in that it possesses a structure: we have seen that from the moment initial functional pairs are constituted we can speak of 'categories' as used by Mac Lane. Their trivial (in this case) character is in fact hommage to the generality of this type of organization which consists of a set of objects with its functions such that associativity exists at the points where composition is possible (without being everywhere applicable) with identities determined by order (or the direction of arrows). This is the fundamental structure which seems to emerge as early as the preoperatory levels.

The most interesting fact about this primitive logic is that it remains essentially qualitative, with a marked dominance of intension over extension, since the latter is not yet regulated, i.e. quantified. This is the basis for the precociousness of identity relations (an essential result not dealt with in this work but to be treated in volume XXIV of this series) due to the coordinator I, in contrast to the notions of conservation, all of which comprise quantification except for the scheme of object permanence which derives precisely from identification.

Although such a relatively coherent logic is very valuable, it has a major flaw when compared to the total logic of operations, for it represents, so to speak, only half a logic, due to the fact that it is oriented according to a given order and still lacks reversibility. Chapter 2, as we have already emphasized, shows this quite clearly since the 'applications' carried out by the subject are one-way, and the succession of genetic stages conforms to the theoretical progression which starts from the notion of application as an oriented correspondence. Inverse correspondences will not come into play (one to many instead of many to one) until the subject has passed from functions to operations and thanks to their reversibility, has constructed hierarchical inclusions.

For lack of this operatory reversibility, the logic of constitutive functions thus neglects inclusions which brings us back to its qualitative nature, since it is thanks to reversible inclusions that operations result in quantifications (from the extension of classes and relations to the construction of number). As soon as operations are elaborated, the incomplete logic of constitutive functions discovers (if we can so express it again) its second half thanks to reversibility, whence the elaboration of groupings or structures of quantified inclusions.

II. – Once the initial coordinators have been completed through operations, and even in certain respects transformed into reversible operators, functions emerge as 'constituted functions' and are then diversified without limit.

We have been able to distinguish two forms of this elaboration. One is still elementary (by comparison to the second level of functions) but it no doubt subsequently plays an essential formatory role, i.e. the constitution of proportionality, to which we will not return. In contrast, the other is fecund without limit, i.e. the construction of functions of functions, through the discovery of variations of variations. From the draining of the liquids of

Chapter 7 or the transformations of squares into rectangles dealt with in Chapter 8, the subject isolates functional mechanisms which derive from the differential calculus and whose schematic or qualitative aspect he has already grasped.

How can the fact that logico-mathematical functions are limitless in number be explained, in contrast to the small number of operations which made possible their elaboration from constitutive functions? This is no more than another version of a question which could have been posed about constitutive functions with regard to the diversity of these original functions when compared to the relative scarcity of the elementary coordinators or combinators.

We could propose two different answers. Either we attribute this fecundity to the contribution of the physical world since functions like actions constitute the common source of causality and of operations; or we invoke a constructivism inherent in operations and parallel (without resulting from them) to the unlimited sequence of 'productions' subsumed under physical causality. The considerations of §3 of this chapter point towards the second solution. It goes without saying that there is a parallelism between operatory constructivity and causal productivity without either one being reducible to the other.[13] This begins when the 'physical coordinators', W', I', (which correspond on the causal plane to the logical combinators W, I, etc.) are constituted on the basis of the dual causal and operation-forming nature of actions, and when causality brings into play the operators 'attributed' to objects corresponding to the operations of the subject.

The functions constituted by means of operations are constructed without limit thanks to a specific process of abstraction. The proportions and variations of variations studied here, although minor examples of the results of this process, are susceptible to unlimited generalization: given certain relationships bearing on the variations of a system, proportions are no more than the relationships among these relationships. Furthermore, given the variations and the differences which they engender, the variations of variations are only analogous functions bearing on the differences themselves. Thus from a system of order n we can draw a system of order $n + 1$ by applying new functions to its own covariations, taking them as objects and isolating the higher order covariations.

This function-generating process is only a specific example of 'reflective abstraction' by means of which the structures of actions or of operations bearing on elements of a certain level are constantly reflected (in the double sense of reflection onto a new plane and of reconstruction by thought) to a higher level of the scale, in turn becoming objects for the elaboration of new structures. The formatory instruments of this indefinite construction are none other than the coordinators or elementary combinators and then the operators in the sense of reversible operations. The products of this construction are the constitutive functions at the preoperatory levels, and then the constituted functions which become much more numerous as the operatory structures are multiplied. It is therefore natural for the products to be richer than the formatory instruments and the higher the level in this progression, the more the distinction between products and instruments becomes blurred, since each of these can become the object of new elaborations as in the inexhaustible interplay of morphisms. We repeat, however, that while this continuous construction of logico-mathematical functions proceeds by reflective abstraction, this operatory progression does not in any way exclude the parallel passage, in accordance with the simple abstractions proper to physical experience, from observed variations to covariations of various orders and from oriented dependences to causal interdependences.

NOTES

[1] By Jean Piaget.

[2] We will nevertheless preserve the term 'prefunctions' (See Vol. XXIV of *Études*, Chap. II, § 2) to designate elementary functions of a psychomorphic nature, for example, when the movement of a living thing (or a thing mistakingly thought to be living) is a function (or seems to be a function) of a goal or of an intention, etc.

[3] As Grize has noted, an object is not a function of another object. However this could take place when one of the objects is the product of the other, as in generation. But then the objects involved themselves become the variables.

[4] We will designate the physical coordinators by the symbols W', I', C' and B' to distinguish them from W, I, C and B.

[5] However one must not exaggerate the precociousness of this belief of the subject according to which nature repeats itself according to constant regularities. First of all, the fact that a given phenomenon is reproduced qualitatively does not by the same token imply that the same quantitative values are repeated. Several of Vinh-Bang's chapters show on the contrary that the conservation of these values in situations which are identical is only acquired towards age 8. (Therein lies the true difference between a quantitative operatory conservation and a qualitative preoperatory identity.) On the other hand, even on the field of qualitative repetitions, there no doubt exists a level (prefunctional and not only preoperatory) where these repetitions do not yet appear to be evident, and we sometimes find traces of this up to age 7–8. We have, for example, pointed out with Inhelder, B. (*The Growth of Logical Thinking from Childhood to Adolescence*) that there are cases where the subject expects that a body which does not float on water will subsequently remain there; where an object rolling down an inclined plane does not always reach the same terminal point; where the repetition of a rebound against a wall (with observed equality of the angles of incidence and of reflection) can engender variations whose effect is cumulative; where the mixture of the same chemical substances does not always result in the production of the same color (here again the

effect of the modification is cumulative); etc. Perhaps what is involved in these cases is a quantitative non-conservation, but it does seem that, according to the complexity of the phenomena, the belief in the qualitative repetition itself is not always as precocious as it is with respect to sufficiently known observables.

[6] We designate these laws or physical functions by the symbol f' for reasons which will soon become evident.

[7] See L'épistémologie du temps, Vol. XX of Études d'épistémologie génétique, 1966, pp. 24–26 and 137–148. As regards time, it might be interesting to recall that Papert and Voyat (Vol. XXI of Études, Chap. II, 5F) showed that 'the discovery of the functional laws linking space-speed and time by far precedes success in problems of hierarchical inclusions of durations of displacements.' Thus we find again the late emergence of reversible inclusions as opposed to oriented functions.

[8] Here the proportionality would be expressed by saying that the large blue flower is to the small blue one as the large red one is to the small red one.

[9] On the other hand, while we do not find the INRC group from age 7–8 (because it comprises a propositional combinatory logic) we observe from this level on multiplicative matrices ($\alpha\beta$; α not β; not α and β; not α nor β), a Klein group which makes it possible to pass from one box to any of the other four.

[10] See Etudes d'épistémologie génétique, Vol. III.

[11] As regards surface areas as such, see the article by Vinh Bang in Vol. XIX of Études (Spatial Conservations).

[12] Which does not in any way preclude, in certain cases, that a mathematical function would imitate or reconstruct a function which was originally discovered physically, in the same way that geometry can deductively reconstruct a physico-spatial configuration.

INDEX

A

Abstraction: 88, 142, 146, 168
  simple, 96, 145, 169–170
  reflective, 97, 146, 168–170, 174,
    194–195
  definition by, 142
Accommodation, 39, 171
Actions: 94, 164, 168, 175, 180, 182–183
  attributed to objects, 167, 176, 179
  causal and operatory, 3, 49
  coordinations of, 30, 75, 88, 95–96,
    168–170
  instrumental, 6, 49, 149
  interiorization of, 7, 170
  inversion of, 62
  schemes of, 3, 12, 15–16, 22, 24, 28,
    30, 46, 74, 88, 95, 170, 182
Addition: 160
  cardinal and ordinal, 37
  ordinal, 171, 187
Algebra: 97, 152
  functional, 149
  Boolean, 156–157
Anticipation, 5, 19–20, 22, 51, 85–86,
  91, 93, 127
Apostel, L., 150–151, 161, 164n, 165n
Applications, 21, 28, 146, 170, 172–174,
  190, 192
  as simplest functions, 13, 15, 20, 23,
    34, 39–40, 46, 58, 74, 87, 130,
    162, 167, 169, 182
  canonical, 158
  defining a quotient-set, 27–29, 151,
    181
  functional, 19, 27, 40
  mathematical, 141, 146
  one-way, 25, 29
  oriented (ordered), 37, 39, 181, 190,
    192
Aristotle, 143
Arithmetic operations, 145
Assimilation: 23, 28, 40, 57, 182
  functional, 171

of actions (primacy of), 88
  recognitory, 33, 172
  schemes of, 3, 39, 95, 171
  of structures, 87
Associativity, 177, 192
Asymmetric relations, 84, 159, 185

B

Bernouilli, J., 145
Bijection: 22, 67, 155
  functions of, 34
Bimorphism, 184
Block, L., 145, 164n
Boolean: algebra, 156–157
  lattices, 152, 157
  structures, 149
Bourbaki, N., 163
Bradwardine, 143
Brandt, 147, 164n, 170
Bresson, 149
Bunge, 159, 165n

C

Calculation, 77, 81
Calculus: infinitesimal, 96
  predicate, 142
Cantor, 143, 164n
Cardinal numbers: 43, 185, 188
  and ordinal numbers, 46, 71, 135
Cardinality, 153, 160
Categories: 4, 170–171, 174, 184–185,
  192
  abstract, 163
  structure of, 8, 12–13, 148
Cauchy, 144
Causal: aspects of actions, 175–176
  circularity, 61, 74, 80
  composition, 89, 96
  comprehension, 55, 168, 179
  consequence, 102n, 136
  explanation, 14, 52, 70, 88, 178, 180
  links, 128, 136, 160, 168

SYNTHESE LIBRARY

Monographs on Epistemology, Logic, Methodology,
Philosophy of Science, Sociology of Science and of Knowledge, and on the
Mathematical Methods of Social and Behavioral Sciences

Managing Editor:
JAAKKO HINTIKKA (Academy of Finland and Stanford University)

Editors:

ROBERT S. COHEN (Boston University)
DONALD DAVIDSON (University of Chicago)
GABRIËL NUCHELMANS (University of Leyden)
WESLEY C. SALMON (University of Arizona)

1. J. M. Bocheński, *A Precis of Mathematical Logic.* 1959, X + 100 pp.
2. P. L. Guiraud, *Problèmes et méthodes de la statistique linguistique.* 1960, VI + 146 pp.
3. Hans Freudenthal (ed.), *The Concept and the Role of the Model in Mathematics and Natural and Social Sciences, Proceedings of a Colloquium held at Utrecht, The Netherlands, January 1960.* 1961, VI + 194 pp.
4. Evert W. Beth, *Formal Methods. An Introduction to Symbolic Logic and the Study of Effective Operations in Arithmetic and Logic.* 1962, XIV + 170 pp.
5. B. H. Kazemier and D. Vuysje (eds.), *Logic and Language. Studies Dedicated to Professor Rudolf Carnap on the Occasion of His Seventieth Birthday.* 1962, VI + 256 pp.
6. Marx W. Wartofsky (ed.), *Proceedings of the Boston Colloquium for the Philosophy of Science, 1961-1962,* Boston Studies in the Philosophy of Science (ed. by Robert S. Cohen and Marx W. Wartofsky), Volume I. 1973, VIII + 212 pp.
7. A. A. Zinov'ev, *Philosophical Problems of Many-Valued Logic.* 1963, XIV + 155 pp.
8. Georges Gurvitch, *The Spectrum of Social Time.* 1964, XXVI + 152 pp.
9. Paul Lorenzen, *Formal Logic.* 1965, VIII + 123 pp.
10. Robert S. Cohen and Marx W. Wartofsky (eds.), *In Honor of Philipp Frank,* Boston Studies in the Philosophy of Science (ed. by Robert S. Cohen and Marx W. Wartofsky), Volume II. 1965, XXXIV + 475 pp.
11. Evert W. Beth, *Mathematical Thought. An Introduction to the Philosophy of Mathematics.* 1965, XII + 208 pp.
12. Evert W. Beth and Jean Piaget, *Mathematical Epistemology and Psychology.* 1966, XII + 326 pp.
13. Guido Küng, *Ontology and the Logistic Analysis of Language. An Enquiry into the Contemporary Views on Universals.* 1967, XI + 210 pp.
14. Robert S. Cohen and Marx W. Wartofsky (eds.), *Proceedings of the Boston Colloquium for the Philosophy of Science 1964-1966, in Memory of Norwood Russell Hanson,* Boston Studies in the Philosophy of Science (ed. by Robert S. Cohen and Marx W. Wartofsky), Volume III. 1967, XLIX + 489 pp.

15. C. D. Broad, *Induction, Probability, and Causation. Selected Papers*. 1968, XI + 296 pp.
16. Günther Patzig, *Aristotle's Theory of the Syllogism. A Logical-Philosophical Study of Book A of the Prior Analytics*. 1968, XVII + 215 pp.
17. Nicholas Rescher, *Topics in Philosophical Logic*. 1968, XIV + 347 pp.
18. Robert S. Cohen and Marx W. Wartofsky (eds.), *Proceedings of the Boston Colloquium for the Philosophy of Science 1966-1968*, Boston Studies in the Philosophy of Science (ed. by Robert S. Cohen and Marx W. Wartofsky), Volume IV. 1969, VIII + 537 pp.
19. Robert S. Cohen and Marx W. Wartofsky (eds.), *Proceedings of the Boston Colloquium for the Philosophy of Science 1966-1968*, Boston Studies in the Philosophy of Science (ed. by Robert S. Cohen and Marx W. Wartofsky), Volume V. 1969, VIII + 482 pp.
20. J.W. Davis, D. J. Hockney, and W. K. Wilson (eds.), *Philosophical Logic*. 1969, VIII + 277 pp.
21. D. Davidson and J. Hintikka (eds.), *Words and Objections: Essays on the Work of W.V. Quine*. 1969, VIII + 366 pp.
22. Patrick Suppes, *Studies in the Methodology and Foundations of Science. Selected Papers from 1911 to 1969*. 1969, XII + 473 pp.
23. Jaakko Hintikka, *Models for Modalities. Selected Essays*. 1969, IX + 220 pp.
24. Nicholas Rescher *et al.* (eds.), *Essays in Honor of Carl G. Hempel. A Tribute on the Occasion of His Sixty-Fifth Birthday*. 1969, VII + 272 pp.
25. P. V. Tavanec (ed.), *Problems of the Logic of Scientific Knowledge*. 1969, XII + 429 pp.
26. Marshall Swain (ed.), *Induction, Acceptance, and Rational Belief*. 1970, VII + 232 pp.
27. Robert S. Cohen and Raymond J. Seeger (eds.), *Ernst Mach: Physicist and Philosopher*, Boston Studies in the Philosophy of Science (ed. by Robert S. Cohen and Marx W. Wartofsky), Volume VI. 1970, VIII + 295 pp.
28. Jaakko Hintikka and Patrick Suppes, *Information and Inference*. 1970, X + 336 pp.
29. Karel Lambert, *Philosophical Problems in Logic. Some Recent Developments*. 1970, VII + 176 pp.
30. Rolf A. Eberle, *Nominalistic Systems*. 1970, IX + 217 pp.
31. Paul Weingartner and Gerhard Zecha (eds.), *Induction, Physics, and Ethics: Proceedings and Discussions of the 1968 Salzburg Colloquium in the Philosophy of Science*. 1970, X + 382 pp.
32. Evert W. Beth, *Aspects of Modern Logic*. 1970, XI + 176 pp.
33. Risto Hilpinen (ed.), *Deontic Logic: Introductory and Systematic Readings*. 1971, VII + 182 pp.
34. Jean-Louis Krivine, *Introduction to Axiomatic Set Theory*. 1971, VII + 98 pp.
35. Joseph D. Sneed, *The Logical Structure of Mathematical Physics*. 1971, XV + 311 pp.
36. Carl R. Kordig, *The Justification of Scientific Change*. 1971, XIV + 119 pp.
37. Milič Čapek, *Bergson and Modern Physics*, Boston Studies in the Philosophy of Science (ed. by Robert S. Cohen and Marx W. Wartofsky), Volume VII. 1971, XV + 414 pp.

38. Norwood Russell Hanson, *What I Do Not Believe, and Other Essays* (ed. by Stephen Toulmin and Harry Woolf), 1971, XII + 390 pp.
39. Roger C. Buck and Robert S. Cohen (eds.), *PSA 1970. In Memory of Rudolf Carnap*, Boston Studies in the Philosophy of Science (ed. by Robert S. Cohen and Marx W. Wartofsky), Volume VIII. 1971, LXVI + 615 pp. Also available as paperback.
40. Donald Davidson and Gilbert Harman (eds.), *Semantics of Natural Language.* 1972, X + 769 pp. Also available as paperback.
41. Yehoshua Bar-Hillel (ed.), *Pragmatics of Natural Languages.* 1971, VII + 231 pp.
42. Sören Stenlund, *Combinators, λ-Terms and Proof Theory.* 1972, 184 pp.
43. Martin Strauss, *Modern Physics and Its Philosophy. Selected Papers in the Logic, History, and Philosophy of Science.* 1972, X + 297 pp.
44. Mario Bunge, *Method, Model and Matter.* 1973, VII + 196 pp.
45. Mario Bunge, *Philosophy of Physics.* 1973, IX + 248 pp.
46. A. A. Zinov'ev, *Foundations of the Logical Theory of Scientific Knowledge (Complex Logic)*, Boston Studies in the Philosophy of Science (ed. by Robert S. Cohen and Marx W. Wartofsky), Volume IX. Revised and enlarged English edition with an appendix, by G. A. Smirnov, E. A. Sidorenka, A. M. Fedina, and L. A. Bobrova. 1973, XXII + 301 pp. Also available as paperback.
47. Ladislav Tondl, *Scientific Procedures*, Boston Studies in the Philosophy of Science (ed. by Robert S. Cohen and Marx W. Wartofsky), Volume X. 1973, XII + 268 pp. Also available as paperback.
48. Norwood Russell Hanson, *Constellations and Conjectures* (ed. by Willard C. Humphreys, Jr.). 1973, X + 282 pp.
49. K. J. J. Hintikka, J. M. E. Moravcsik, and P. Suppes (eds.), *Approaches to Natural Language. Proceedings of the 1970 Stanford Workshop on Grammar and Semantics.* 1973, VIII + 526 pp. Also available as paperback.
50. Mario Bunge (ed.), *Exact Philosophy – Problems, Tools, and Goals.* 1973, X + 214 pp.
51. Radu J. Bogdan and Ilkka Niiniluoto (eds.), *Logic, Language, and Probability. A Selection of Papers Contributed to Sections IV, VI, and XI of the Fourth International Congress for Logic, Methodology, and Philosophy of Science, Bucharest, September 1971.* 1973, X + 323 pp.
52. Glenn Pearce and Patrick Maynard (eds.), *Conceptual Chance.* 1973, XII + 282 pp.
53. Ilkka Niiniluoto and Raimo Tuomela, *Theoretical Concepts and Hypothetico-Inductive Inference.* 1973, VII + 264 pp.
54. Roland Fraïssé, *Course of Mathematical Logic* – Volume 1: *Relation and Logical Formula.* 1973, XVI + 186 pp. Also available as paperback.
55. Adolf Grünbaum, *Philosophical Problems of Space and Time.* Second, enlarged edition, Boston Studies in the Philosophy of Science (ed. by Robert S. Cohen and Marx W. Wartofsky), Volume XII. 1973, XXIII + 884 pp. Also available as paperback.
56. Patrick Suppes (ed.), *Space, Time, and Geometry.* 1973, XI + 424 pp.
57. Hans Kelsen, *Essays in Legal and Moral Philosophy*, selected and introduced by Ota Weinberger. 1973, XXVIII + 300 pp.
58. R. J. Seeger and Robert S. Cohen (eds.), *Philosophical Foundations of Science. Proceedings of an AAAS Program, 1969*, Boston Studies in the Philosophy of

Science (ed. by Robert S. Cohen and Marx W. Wartofsky), Volume XI. 1974, X + 545 pp. Also available as paperback.

59. Robert S. Cohen and Marx W. Wartofsky (eds.), *Logical and Epistemological Studies in Contemporary Physics*, Boston Studies in the Philosophy of Science (ed. by Robert S. Cohen and Marx W. Wartofsky), Volume XIII. 1973, VIII + 462 pp. Also available as paperback.

60. Robert S. Cohen and Marx W. Wartofsky (eds.), *Methodological and Historical Essays in the Natural and Social Sciences. Proceedings of the Boston Colloquium for the Philosophy of Science, 1969-1972,* Boston Studies in the Philosophy of Science (ed. by Robert S. Cohen and Marx W. Wartofsky), Volume XIV. 1974, VIII + 405 pp. Also available as paperback.

61. Robert S. Cohen, J. J. Stachel and Marx W. Wartofsky (eds.), *For Dirk Struik. Scientific, Historical and Political Essays in Honor of Dirk J. Struik*, Boston Studies in the Philosophy of Science (ed. by Robert S. Cohen and Marx W. Wartofsky), Volume XV. 1974, XXVII + 652 pp. Also available as paperback.

62. Kazimierz Ajdukiewicz, *Pragmatic Logic*, transl. from the Polish by Olgierd Wojtasiewicz. 1974, XV + 460 pp.

63. Sören Stenlund (ed.), *Logical Theory and Semantic Analysis. Essays Dedicated to Stig Kanger on His Fiftieth Birthday*. 1974, V + 217 pp.

64. Kenneth F. Schaffner and Robert S. Cohen (eds.), *Proceedings of the 1972 Biennial Meeting, Philosophy of Science Association*, Boston Studies in the Philosophy of Science (ed. by Robert S. Cohen and Marx W. Wartofsky), Volume XX. 1974, IX + 444 pp. Also available as paperback.

65. Henry E. Kyburg, Jr., *The Logical Foundations of Statistical Inference*. 1974, IX + 421 pp.

66. Marjorie Grene, *The Understanding of Nature: Essays in the Philosophy of Biology*, Boston Studies in the Philosophy of Science (ed. by Robert S. Cohen and Marx W. Wartofsky), Volume XXIII. 1974, XII + 360 pp. Also available as paperback.

67. Jan M. Broekman, *Structuralism: Moscow, Prague, Paris*. 1974, IX + 117 pp.

68. Norman Geschwind, *Selected Papers on Language and the Brain*, Boston Studies in the Philosophy of Science (ed. by Robert S. Cohen and Marx W. Wartofsky), Volume XVI. 1974, XII + 549 pp. Also available as paperback.

69. Roland Fraïssé, *Course of Mathematical Logic* – Volume 2: *Model Theory*. 1974, XIX + 192 pp.

70. Andrzej Grzegorczyk, *An Outline of Mathematical Logic. Fundamental Results and Notions Explained with All Details*. 1974, X + 596 pp.

71. Franz von Kutschera, *Philosophy of Language*. 1975, VII + 305 pp.

72. Juha Manninen and Raimo Tuomela (eds.), *Essays on Explanation and Understanding. Studies in the Foundations of Humanities and Social Sciences*. 1976, VII + 440 pp.

73. Jaakko Hintikka (ed.), *Rudolf Carnap, Logical Empiricist. Materials and Perspectives*. 1975, LXVIII + 400 pp.

74. Milič Čapek (ed.), *The Concepts of Space and Time. Their Structure and Their Development*, Boston Studies in the Philosophy of Science (ed. by Robert S. Cohen and Marx W. Wartofsky), Volume XXII. 1976, LVI + 570 pp. Also available as paperback.

75. Jaakko Hintikka and Unto Remes, *The Method of Analysis. Its Geometrical Origin and Its General Significance,* Boston Studies in the Philosophy of Science (ed. by Robert S. Cohen and Marx W. Wartofsky), Volume XXV. 1974, XVIII + 144 pp. Also available as paperback.
76. John Emery Murdoch and Edith Dudley Sylla, *The Cultural Context of Medieval Learning. Proceedings of the First International Colloquium on Philosophy, Science, and Theology in the Middle Ages – September 1973,* Boston Studies in the Philosophy of Science (ed. by Robert S. Cohen and Marx W. Wartofsky), Volume XXVI. 1975, X + 566 pp. Also available as paperback.
77. Stefan Amsterdamski, *Between Experience and Metaphysics. Philosophical Problems of the Evolution of Science,* Boston Studies in the Philosophy of Science (ed. by Robert S. Cohen and Marx W. Wartofsky), Volume XXXV. 1975, XVIII + 193 pp. Also available as paperback.
78. Patrick Suppes (ed.), *Logic and Probability in Quantum Mechanics.* 1976, XV + 541 pp.
79. H. von Helmholtz, *Epistemological Writings.* (A New Selection Based upon the 1921 Volume edited by Paul Hertz and Moritz Schlick, Newly Translated and Edited by R. S. Cohen and Y. Elkana), Boston Studies in the Philosophy of Science, Volume XXXVII. 1977 (forthcoming).
80. Joseph Agassi, *Science in Flux,* Boston Studies in the Philosophy of Science (ed. by Robert S. Cohen and Marx W. Wartofsky), Volume XXVIII. 1975, XXVI + 553 pp. Also available as paperback.
81. Sandra G. Harding (ed.), *Can Theories Be Refuted? Essays on the Duhem-Quine Thesis.* 1976, XXI + 318 pp. Also available as paperback.
82. Stefan Nowak, *Methodology of Sociological Research: General Problems.* 1977, XVIII + 504 pp. (forthcoming).
83. Jean Piaget, Jean-Blaise Grize, Alina Szeminska, and Vinh Bang, *Epistemology and Psychology of Functions.* 1977 (forthcoming).
84. Marjorie Grene and Everett Mendelsohn (eds.), *Topics in the Philosophy of Biology,* Boston Studies in the Philosophy of Science (ed. by Robert S. Cohen and Marx W. Wartofsky), Volume XXVII. 1976, XIII + 454 pp. Also available as paperback.
85. E. Fischbein, *The Intuitive Sources of Probabilistic Thinking in Children.* 1975, XIII + 204 pp.
86. Ernest W. Adams, *The Logic of Conditionals. An Application of Probability to Deductive Logic.* 1975, XIII + 156 pp.
87. Marian Przełęcki and Ryszard Wójcicki (eds.), *Twenty-Five Years of Logical Methodology in Poland.* 1977, VIII + 803 pp. (forthcoming).
88. J. Topolski, *The Methodology of History.* 1976, X + 673 pp.
89. A. Kasher (ed.), *Language in Focus: Foundations, Methods and Systems. Essays Dedicated to Yehoshua Bar-Hillel,* Boston Studies in the Philosophy of Science (ed. by Robert S. Cohen and Marx W. Wartofsky), Volume XLIII. 1976, XXVIII + 679 pp. Also available as paperback.
90. Jaakko Hintikka, *The Intentions of Intentionality and Other New Models for Modalities.* 1975, XVIII + 262 pp. Also available as paperback.
91. Wolfgang Stegmüller, *Collected Papers on Epistemology, Philosophy of Science and History of Philosophy,* 2 Volumes, 1977 (forthcoming).

92. Dov M. Gabbay, *Investigations in Modal and Tense Logics with Applications to Problems in Philosophy and Linguistics*. 1976, XI + 306 pp.
93. Radu J. Bogdan, *Local Induction*. 1976, XIV + 340 pp.
94. Stefan Nowak, *Understanding and Prediction: Essays in the Methodology of Social and Behavioral Theories*. 1976, XIX + 482 pp.
95. Peter Mittelstaedt, *Philosophical Problems of Modern Physics*, Boston Studies in the Philosophy of Science (ed. by Robert S. Cohen and Marx W. Wartofsky), Volume XVIII. 1976, X + 211 pp. Also available as paperback.
96. Gerald Holton and William Blanpied (eds.), *Science and Its Public: The Changing Relationship*, Boston Studies in the Philosophy of Science (ed. by Robert S. Cohen and Marx W. Wartofsky), Volume XXXIII. 1976, XXV + 289 pp. Also available as paperback.
97. Myles Brand and Douglas Walton (eds.), *Action Theory. Proceedings of the Winnipeg Conference on Human Action, Held at Winnipeg, Manitoba, Canada, 9-11 May 1975*. 1976, VI + 345 pp.
98. Risto Hilpinen, *Knowledge and Rational Belief*. 1978 (forthcoming).
99. R. S. Cohen, P. K. Feyerabend, and M. W. Wartofsky (eds.), *Essays in Memory of Imre Lakatos*, Boston Studies in the Philosophy of Science (ed. by Robert S. Cohen and Marx W. Wartofsky), Volume XXXIX. 1976, XI + 762 pp. Also available as paperback.
100. R. S. Cohen and J. Stachel (eds.), *Leon Rosenfeld, Selected Papers*. Boston Studies in the Philosophy of Science (ed. by Robert S. Cohen and Marx W. Wartofsky), Volume XXI. 1977 (forthcoming).
101. R. S. Cohen, C. A. Hooker, A. C. Michalos, and J. W. van Evra (eds.), *PSA 1974: Proceedings of the 1974 Biennial Meeting of the Philosophy of Science Association*, Boston Studies in the Philosophy of Science (ed. by Robert S. Cohen and Marx W. Wartofsky), Volume XXXII. 1976, XIII + 734 pp. Also available as paperback.
102. Yehuda Fried and Joseph Agassi, *Paranoia: A Study in Diagnosis*, Boston Studies in the Philosophy of Science (ed. by Robert S. Cohen and Marx W. Wartofsky), Volume L. 1976, XV + 212 pp. Also available as paperback.
103. Marian Przełęcki, Klemens Szaniawski, and Ryszard Wójcicki (eds.), *Formal Methods in the Methodology of Empirical Sciences*. 1976, 455 pp.
104. John M. Vickers, *Belief and Probability*. 1976, VIII + 202 pp.
105. Kurt H. Wolff, *Surrender and Catch: Experience and Inquiry Today*, Boston Studies in the Philosophy of Science (ed. by Robert S. Cohen and Marx W. Wartofsky), Volume LI. 1976, XII + 410 pp. Also available as paperback.
106. Karel Kosík, *Dialectics of the Concrete*, Boston Studies in the Philosophy of Science (ed. by Robert S. Cohen and Marx W. Wartofsky), Volume LII. 1976, VIII + 158 pp. Also available as paperback.
107. Nelson Goodman, *The Structure of Appearance*, Boston Studies in the Philosophy of Science (ed. by Robert S. Cohen and Marx W. Wartofsky), Volume LIII. 1977 (forthcoming).
108. Jerzy Giedymin (ed.), *Kazimierz Ajdukiewicz: Scientific World-Perspective and Other Essays, 1931–1963*. 1977 (forthcoming).
109. Robert L. Causey, *Unity of Science*. 1977, VIII+185 pp.
110. Richard Grandy, *Advanced Logic for Applications*. 1977 (forthcoming).